HIGH POWER
SWITCHING

HIGH POWER SWITCHING

Ihor Vitkovitsky

Plasma Physics Division
Naval Research Laboratory

VAN NOSTRAND REINHOLD COMPANY
New York

Copyright © 1987 by Van Nostrand Reinhold Company Inc.

Library of Congress Catalog Card Number 86-28126

ISBN 0-442-29067-5

Printed in the United States of America

Van Nostrand Reinhold Company Inc.
115 Fifth Avenue
New York, New York 10003

Van Nostrand Reinhold Company Limited
Molly Millars Lane
Wokingham, Berkshire RG11 2PY, England

Van Nostrand Reinhold
480 La Trobe Street
Melbourne, Victoria 3000, Australia

Macmillan of Canada
Division of Canada Publishing Corporation
164 Commander Boulevard
Agincourt, Ontario M1S 3C7, Canada

16 15 14 13 12 11 10 9 8 7 6 5 4 3 2 1

Library of Congress Cataloging-in-Publication Data

Vitkovitsky, Ihor M., 1932–
    High power switching.

    Includes index.
    1. Electric switchgear. 2. Electromagnetic pulse.
I. Title.
TK2828.V58   1987   621.31′7   86-28126
ISBN 0-442-29067-5

To my wife, Tamara

Preface

Pulsed power technology is an area of rapid development propelled by potential benefits expected from thermonuclear fusion energy sources, from industrial and military applications of larger lasers, from the ever-increasing sophistication of electric power distribution (and consequently more demanding protection requirement of power lines and equipment), and from a variety of new specialized needs. This technology deals with the generation of very high power electromagnetic pulses, as distinct from the continuous production of power, and with the coupling of pulses to loads. An important segment of pulsed power technology is high power switching. This book deals with switching technology used in applications where pulser output exceeds 10^9 watts. In most cases, such output is a monopulse with current and voltage amplitudes ranging from kiloamperes and kilovolts to megamperes and megavolts. More difficult applications, which use trains or bursts of pulses to drive lasers, microwave sources, and other high power devices, have appeared recently, leading to attempts to extend high power capability to repetitively operated switches. High power switching plays a key role in advancing new technologies, such as thermonuclear energy sources in the civilian sector and technologies such as directed energy beam weapons, nuclear weapon effects simulators, and ultra-high velocity guns in the military sector. Specifically, in each of these sectors there is a need for significant improvements in closing switches and for continuing the development of new opening and repetitively operated switches that would lead to less costly and more reliable, efficient, and compact pulse power systems.

This book is intended to serve as a source of information on existing types of high power switches, and on new switching concepts currently being developed or those that are in speculative stages. The book consists of two major chapter groups in addition to the introductory chapters and Appendix. One group (Chapters 3 through 6) deals with closing switches, and the other group (Chapters 7 through 10) deals with opening switches and includes a discussion of repetitively operated switches. Typical switch applications are illustrated where appropriate. The first two chapters consist of a tutorial discussion of general switching principles and of circuits employing high power switches. The discussion is limited to idealized circuits to allow the reader to relate switch performance to that of the entire pulser without being burdened with the intricacies of other considerations. The last chapter conveys peripheral information related to electromagnetic noise problems and considers the important safety aspects of switch operation. The book includes basic physical principles, engineering data, and scaling laws that permit the reader to assess both the applicability of a given switch in specific areas of interest and the risks versus potential pay-off associated with future development of given switch types or pulser systems. The chapters contain many references to sources for additional information.

In concept, *High Power Switching* is a reference surveying many types of switches and discussing their operating principles and performance. In numerous instances, examples of applications have been provided to demonstrate their potential, as well as to emphasize the interdependence of the switches, the pulser system, and the load. The field of high power switches continues to develop as new applications of pulsed power arise and new concepts emerge to overcome the older technological bounds. Therefore, throughout the book, potential new directions for development are highlighted.

At this time (1987), no convenient, accessible monograph is available that surveys both closing and opening switches. Some valuable material on closing switches has been collected in such works as those of V. S. Komelkov[1] and T. Burkes[2], which are limited, not easily accessible editions. *Pulse Power Notes Series*[3] should be noted by those interested in switches as well as in other components of high power pulsed sources. None of the published compendia includes an extensive exposition of opening switch concepts and technology; limited discussion of opening switches appears in publications such as Mesyats[4]. Relevant material on both types of switches is available in books dealing with pulse power generation and application. Some texts deal with specialized applications of switches, for example, *Current Interruption in High Voltage Networks*[5], edited by K. Ragaller, dealing with electric power line circuit-breakers. And, of course, there are published proceedings of many conferences, workshops and special-issue periodical journals. These however, are not widely available and usually take for granted the specialized technical background of its audience. This book attempts to remedy, in part, this situation and devotes the entire second group of chapters to opening switch technology, including its most recent embodiments. Both one-shot switches that are destroyed as a result of the opening action and those that can be reused are described.

The author and his colleagues at the Naval Research Laboratory were key contributors to the development of some of the switches discussed in this volume. These colleagues have generously provided some written material for Chapters 5, 8 and 9. Other switch types have been developed elsewhere, and in many cases, illustrations and performance data have been included in this book with permission of several publishers, for which I am very grateful.

I am also very grateful to several colleagues, known experts in pulse power technology, for critical reviews of the various chapters in the book. I would like to thank R. E. Pechacek for consultation regarding the material in Chapter 3 and for the review by D. Hinshelwood that raised significant points subsequently incorporated into Chapter 4. In reviewing Chapter 5, J. D. Shipman wrote in his comments to me that "it would have been very nice to have had material such as this available when we started out." This comment was extremely encouraging and helpful in my perserverance with the manuscript. R. A. Miller provided an invaluable critique of Chapter 6, as well as some material for this chapter. R. Dethlefsen reviewed Chapter 7 from the viewpoint of an expert on electrical power switchgear. W. H. Lupton's knowledge of exploding wires and fuses led to many discussions in preparation of Chapter 8. R. J. Commisso supplied very useful material and discussion for Chapter 9 and R. J. Meger ably reviewed this chapter. Finally, P. J. Turchi provided the critique and made suggestions for improving Chapter 10. The overall review of the book and the continuing encouragement with this project by Prof. W. J. Sargeant of the State University of New York at Buffalo is greatly appreciated.

I am also greatful to various persons within the Naval Research Laboratory as well as outside for programmatic support. Dr. G. Cooperstein, Dr. S. L. Ossakow, and Dr. T. Coffey are thanked for their support and their enthusiasm for the various pulse power science and technology programs at the Naval Research Laboratory. This support has resulted in many switch developments discussed in this book. Dr. T. Coffey is especially to be thanked for his early recognition of pulse power technology as the key to advances in many fields of national interest and his active support of these programs. Mr. P. Haas and Mr. J. Farber of the Defense Nuclear Agency were the primary supporters of the pulse power programs both at the Naval Research Laboratory and elsewhere over many years. These individuals and many others in various laboratories and institutions had the foresight to allow many of the concepts in switching to come to fruition and those results are reported here. I feel especially indebted to J. C. Martin of the Atomic Weapons Research Establishment in Great Britian for his early perspective on switching technology that stimulated much of the high power switch development reported here. This book draws on the work of many other individuals and their contributions are reflected through the references. Personal contacts with many of them were invaluable in the preparation of the manuscript for *High Power Switching*.

Ihor Vitkovitsky

REFERENCES

1. V. S. Komelkov, "Technology of Large Impulse Currents and Magnetic Fields," translated by Foreign Technology Division, Wright-Patterson AFB, Ohio, Report FTD-MT-24-992-71, from *Tekhnika Bolshikh Impulsnykn Tokov i Magnitnykh Poley,* Atomizdat, Moscow (1970).
2. T. R. Burkes, "A Critical Analysis and Assessment of High Power Switches," Naval Surface Weapon Center (Dahlgren, VA), Report NP 30/78 (1978).
3. C. Baum, ed., "Pulsed Electrical Power Circuit and Electromagnetic System Design Notes, PEP 4-1," Air Force Weapons Laboratory (Albuquerque, NM), Report AFWL-TR-73-166 (1973).
4. G. A. Mesyats, ed., *High Power Nanosecond Pulsed Sources for Electron Acceleration* (in Russian), Nauka Publishers, Novosibirsk, USSR (1974).
5. K. Ragaller, ed., *Current Interruption in High-Voltage Networks,* Plenum, New York (1978).

Contents

HIGH POWER
SWITCHING

1
Introduction

SWITCHES IN TECHNOLOGY

Growth of electric power distribution networks in the first part of the twentieth century stimulated the improvement of network switching, as well as the development of switches required to operate electrical equipment at ever increasing power. It became necessary to develop the technology not only for switchgear to control power distribution, but also for protection against abnormally high power surges, which became imperative in order to protect expensive equipment tied to power lines and to assure that electric power service would be restored in the shortest possible time. While the early electric power networks required protection at low power levels, the post-World War II period witnessed the development of power grids with currents above 100 kA and transmission voltages of 1 MV level[1]. This required switchgear to handle power levels much higher than 10^9 W. Such technology became a base for the development of switches for applications employing single pulses of current. In this book, circuit breakers, used for electric power transmission and their derivatives, developed for pulse power applications, are discussed in Chapter 7. Because of the widespread use and economic importance of switchgear for power distribution and protection, many detailed treatises dealing with the underlying principles, design criteria and applications have been published, for example, Ref. 2. The need for much higher power switches became evident, as their application in various experimental facilities to drive particle accelerators, intense magnetic field generators, and high power radio transmitters was established. P.L. Kapitsa in the early 1920's used the mechanical energy of an electrical generator rotor to produce short circuit current as a means of inducing 50-tesla magnetic fields[3], which for that time were considered extremely high. Such experiments led to the evolution of more sophisticated power generation as well as of switching techniques required for transferring power pulses to the load. One of the first generators specifically constructed for pulsed applications in high energy particle accelerators (synchrotrons) was a large homopolar generator. It was integrated with the new switching design and, tested at a current level of nearly 3 MA at 100 V output[4].

In the second half of the twentieth century, attempts to exploit new concepts for generating electric power based on nuclear fusion in plasmas confined by strong magnetic fields have been initiated. This has led to development of a new class of closing and opening switches capable of extremely fast switching times. To explore the feasibility of controlled thermonuclear burn for electric power generation, it has become necessary to utilize much higher currents than those produced previously and to reach peak current values in times much shorter than could be provided by rotating machinery or batteries. The shift in switch capability toward higher currents and to shorter pulse times becomes, in time, a foundation of the newly developing pulse power technology, leading to investigations of a large variety of very high power switches capable of handling powers approaching levels higher than 10^{12} W. An example of such a switch is described in Chapter 6 (and shown in Fig. 6-7). It combines the design features of low power switches with techniques of mitigating the strong mechanical forces arising in high power switching. Some switching methods, nevertheless, have proved to be difficult to extend to higher powers. One type of switch, the thyratron[5], could not be adapted to handle power at 10^9 W level and therefore, is not included in the content of this book, even though it is included in other reviews which contain some discussion of pulse power switching[5, 6]. The shift in switch parameters to higher currents was recognized by V.S. Komelkov, who then proceeded to publish in 1970 a significant comprehensive survey of the pulsed high current technology[3], discussing its two major elements—the power sources, such as capacitor banks and inductors, and high power switches.

The broadening of thermonuclear fusion concepts to include attempts for production of electric power using inertially confined plasmas[7] heated by very intense lasers, electron, or ion beams made high power, low inductance switching even more critical to the development of an inexhaustible supply of electric power. Rapid pulsing of input power to heat inertially confined plasmas as well as to provide power sources for various other application placed another requirement on switching technology: repetitive operation at very high power levels[8]. This aspect of switch design, the capability to switch repetitively, is discussed throughout this book, with special attention devoted to it in Chapters 4 and 9.

Technology areas other than those already mentioned continue to stimulate development of high power switches. Intense lasers and charged particle beam generators require power sources with output of 10^9 to 10^{12} W and even a higher power regime[9]. Simulation of x-ray and electromagnetic pulses from nuclear weapons also exploits sources in this regime. Simulator sources and their characteristics are listed, for example, in Table 1.1 of Ref. 10. Invariably, the design of high power sources ends up with some part of the system utilizing spark gap switches. This type of switch is most commonly used in high power systems. A large section of this book is, therefore, devoted to various spark gap designs. Gaps with gas, liquid and solid insulation between the electrodes are described in Chapters 3, 5, and 6, respectively. Further applications, such as propulsion of projectiles by strong magnetic fields[11] (e.g., to simulate collisions of meteorites with spacecraft or to produce super-high pressure pulses) constitute other examples of the high power switches, which must handle very large currents in the millions of amperes.

The phenomenon of switching is not limited to man-made devices—it also appears in nature. Lightning discharge represents, in terms of scale, the largest closing switch. It is used by nature to discharge the electrified earth-cloud parallel plate "capacitor" system. Figure 1–1 is a photograph of the natural "switching" occuring in the Albuquerque, New Mexico, area[12].

In the span of three decades, high power switching has become a distinct field of applied physics. As the use of switches has proliferated with the growth of new applications, the literature describing the physics and technology of high power switching has accordingly grown in volume. The most comprehensive list of references related to the pulse power technology, and within it to switching, is the annotated and indexed bibliography edited by

Fig. 1–1. Thunderstorm over Alburquerque, NM, depicts the switching of the electrostatic energy stored between the clouds and the ground.[12] Albuquerque Journal Photo by Brian Walski.

J. Bernesderfer et al.[13] It contains about 2500 full bibliographic citations, original sources, availability, key words and abstracts. Through the use of these indexes—subject matter, personal author and corporate author, with world coverage—the bibliography becomes an indispensable tool for the designer of high power switching and for the developer of pulsed power systems.

CHARACTERIZATION OF SWITCHES

Diverse applications of high power switches requires many types of switches with a broad range of characteristics. These characteristics can be grouped into those relating to electrical capabilities of the switch and those relating to its physical, operational, and other features. The succeeding chapters generally follow this division. Switch voltage characteristics are

- Hold-off, or stand-off voltage.
- Voltage drop across the switch impedance (during conduction).
- Rate of rise or drop of voltage across the switch for opening and closing switches, repetitively.
- Prefire or switching below nominal switch hold-off voltage.
- Trigger voltage, or secondary voltage pulse necessary to initiate switching.

In addition to voltage characteristics, there are several characteristics of switches related to the current conducted by the switch. These are

- Maximum current that can pass through the switch without damage.
- Maximum charge (or the time integral of current) that can pass through the switch without damage.
- Rate of rise of current allowed by the switch impedance.

The product of the current through the switch and hold-off voltage, even though not necessarily occurring at the same time, provides the customary definition of power of the pulse that the switch handles. This definition of *power* handled by the switch is used in this book. The definition applies to closing switches (with the voltage taken to be that before switching and the current peak occuring after the voltage across the switch collapses) and to opening switches (where the current is that during conduction before the opening of the switch, and the voltage appears subsequent to the current drop in the switch).

The *energy* handled by the switch is defined more rigorously, as the time integral of the product of current and voltage during switching, i.e., while the current and voltage of the switch are changing due to switching action. This time interval and other characteristic intervals are

- Time of current conduction through the switch.
- Time of voltage hold-off.
- Opening time.
- Closing time.
- Switching delay, or the interval between trigger signal and switch closing or opening.
- Jitter time, or the deviation from nominal switching delay time.
- Recovery time, or time required by the switch to become ready for next operation.
- Repetition rate (with the maximum rate being the inverse of the recovery rate).

Obviously not all the factors listed above are of consequence to each switch operation. Depending on the switch application, some of the following characteristics can also be significant and frequently dominate the choice of switches or determine their design:

- Lifetime, or the number of switching operations before failure.
- Reliability.
- Maintainability.
- Fault modes, or ways in which the switch fails.
- Ease of installation.
- Weight and volume.
- Cost.

Other characteristics also become important in those circumstances where large energy transfer through the switch occurs. Explosive forces arising from rapid deposition of electrical energy due to ohmic losses in the switch, or from large magnetic currents passing through the switch also require attention. For example, special methods for moderating such forces in opening switches are discussed in some detail. In other circumstances, switch operation influences the design of the circuit served by such switches, as in the case of spark gaps for shaping power flow in large pulse lines, where the pulse line and the switch form an integral component. The interaction between the circuit and the switch is emphasized in discussion of liquid dielectric spark gaps in Chapter 6, i.e., in those power systems where the high dielectric constant of the liquid insulator significantly reduces the electromagnetic wave transit time with resultant shrinking of the system dimensions.

Many of the switch characteristics described above are assigned symbols for rigorous discussion of the physical phenomena, for prescribing the circuit parameters and for characterizing the switch performance. The symbols for more common quantities such as current, voltage, resistance and the intrinsic quantities, such as resistivity or density, are the same throughout the book. The symbols for more special characteristics, such as time characteristics or dimensions, relating to given switch or associated with the specific circuit, are defined in each chapter for the specific cases.

The units are mks in most cases. In a few instances, other units, specifically indicated, are used to maintain the simplicity of commonly used formulae.

PHYSICAL PROCESSES

High power switch technology employs a wide spectrum of physical principles and engineering practices. Although most of the phenomena associated with switching are well understood, there are also areas which are poorly understood and require further studies. To supplement the lack of understanding there has accumulated a large body of empirical data and scaling laws for relatively narrow operating regimes to help the switch designer make appropriate choices for optimizing pulse power systems. Some of the newer types of switches, such as those discussed in Chapter 10, which use phase transitions in solids to change the bulk resistivity by many orders of magnitude have potential for development of very practical devices once the physics determining the material behavior is fully understood and the technology of forming materials with suitable properties is mastered.

The dominant physical phenomena which govern the switch performance are determined first and foremost by the dielectric used as the insulator in the switch. The dielectric between the switch electrodes can be a solid material, liquid, or gas at high, atmospheric, or very low pressure. Ionized gases at low pressures (plasmas) also are employed in switches, with the use of magnetic fields to turn off conduction, rendering such plasmas as effective insulators with hold-off fields of several MV/cm. The closing switch action in solids, liquids, and gases—formation of a discharge—can be initiated by a variety of means, including self-breakdown or triggering. The opening switch can interrupt the current flow with a large variety of methods which in one way or another manage to change resistance of the switch from the initial low value to a final high value.

The most significant characteristics governing the performance of the opening switches are the insulation properties, transition rate from insulation (i.e., high resistivity) to conduction and vice versa, and current density at which such transition can occur reversibly. The most significant factor in determining the switching range is the nature of conduction in the switch. Both types of switches can be divided into two groups, according to the type of conduction which they employ during the conduction phase. One type of switch employs channel conduction, associated with highly localized flow of current and produces high energy density; the resulting heating leads to formation of ionized gas or plasma. Volume conduction, where the current to be switched is made to flow over a large cross-sectional area, prevents formation of very high energy density, leading to more controllable transition from one resistance state to another state. Chapters 3, 4, 5, 6, 7, and 8 deal mainly with switches utilizing channel discharges. Chapters 9 and 10 describe switches based on volume conduction.

The high energy density of channel discharge switches also leads to significant pressures and explosive forces. This requires that switch structural integrity must be considered. Volume discharge switches greatly reduce or eliminate the problems caused by strong mechanical impulses and therefore are attractive for high power switch designs. However, because the price for volume conduction may be high, in terms of switch complexity, its practicality or reliability (for example, complex ionization sources may be needed to assure volume conduction), volume discharge switches are not used frequently; thus, much effort has been devoted to finding methods to reduce or redistribute the impulse (for example, by subdividing the discharge into multiple channels). The pressure impulses generated by the channel discharges further complicate the design of some switches, because it becomes necessary to consider the resulting dynamic phenomena, such as shocks, since they can alter significantly the insulating properties of the regions subjected to the electric stress by the potential applied across the switch.

Less encompassing, but nevertheless important phenomena in many switches are those associated with electrode surface physics and with various predischarge effects. Surface control in low pressure switches is the main factor in determining reliable switch performance. Electrode erosion and deposition of the products resulting from the switching action determines the subsequent switch performance and plays a very important role in determining switch lifetime.

The performance of switches is also related to circuits in which they are employed. High power switches are typically employed in circuits containing energy storage elements. The characteristics of such elements determine the current through the switch, voltage across the switch and their duration, as well as mechanical characteristics. Chapter 2 discusses basic circuits with energy storage elements used as current or voltage sources and the role of switches in transforming the stored energy into an appropriate pulse shape and its subsequent transfer to the load.

In high power applications, these circuits are relatively uncomplicated, but geometric and mechanical considerations influence their performance significantly. This situation arises from the need to package large amounts of stored energy near the load. In the limit, the maximum power transfer is determined by the transit time for the electromagnetic pulse from storage to load; that is, it depends on the ratio of the linear dimension of the storage system to the velocity of the electromagnetic wave in the given storage medium. The maximum power in such a limiting case is the energy stored divided by the transit time.

For pulsed power systems not limited by such a fundamental constraint, it often makes sense to package the stored energy into a minimum volume on the basis of economics or because of practical considerations such as portability. High energy density electrical storage systems are discussed, for example, by Knoepfel[14] in conjunction with the generation of very high magnetic fields. To rank the storage systems according to maximum power which they can deliver, assuming that it is not limited by the switch connecting the storage system with the load, the energy density of the storage systems is considered.

The energy densities of such storage systems range over many orders of magnitude. Capacitor energy density varies from 10^4 J/m^3 for more common systems[14] to 10^6 J/m^3 in specialized products[15]. The energy density of rotating machinery is of the order of 5×10^8 J/m^3 in flywheels[14] and 5×10^6 J/m^3 in complete inertial-inductive packages, such as self-excited (pulsed) homopolar generator[16]. The highest energy density, excluding nuclear explosives, is associated with chemical explosives[14]. It is about 10^{10} J/m^3 and an order of magnitude less in terms of the electrical energy generated by using explosives in production of pulsed electrical output for rapid compression of magnetic fields (i.e., magneto-cumulative generators)[14].

It turns out that often switches do limit the output power of an energy storage system. One of the factors of high power switch design, emphasized in Chapter 2 in relation to efficiency of energy transfer from the storage element to the load, is the need to minimize switch inductance. Because pulsed power system inductance represents a spatial distribution of conductors in the circuit, including those assocaited with the storage system and with the switch, the requirement for low inductance is equivalent to the above statements regarding the need for high energy density in storage elements and in the switches. Thus, the switch geometry must be such that it retains the short distance between the energy store and the load. This suggests that switch dimensions should be minimized, which in turn implies high energy density in the switch. Figure 1–2 shows schematically the energy density associated with operation of switches suitable for high power applications. The range of switch energy densities is about the same as for energy storage.

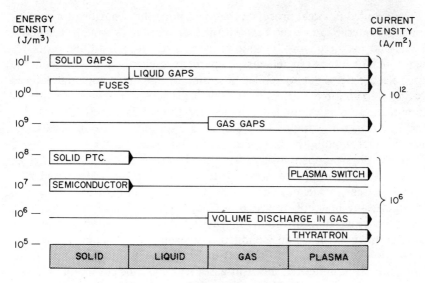

Fig. 1–2. Chart showing main categories of switches and the state of matter used for switching. Some switch types operate in one state only, e.g., the positive temperature coefficient (PTC) bulk switches employ a solid phase material both during conduction and during current interruption. Plasma switches employ the plasma state only. Most of the other switch categories include more than one state of matter. The direction of phase transitions is shown by the arrows. The typical switch energy density is shown at left and the range of current density in the switch is shown at the right.

The energy density assignments in Fig. 1–2 are associated with the part of the switch undergoing the change in resistance which produces the switching, namely, a drastic change in current in some branch of a circuit. Thus, in a simple spark gap, the energy density of 10^9 J/m^3 is associated with the arc channel. In case of a volume discharge in the gas, the product of current density and the electric field sustained by the switch over the switching time yields lower energy density, in the range of 10^6 J/m^3. Figure 1–2 also shows the phase of the matter employed by each type of switch as well as transitions from one state to another needed to change resistance in some of the switch types. Thus, switches based on use of fuses start out as a solid (during conduction), make a transition to a liquid state, then vaporize due to ohmic heating. The transition to vapor is associated with the large increase in resistance, i.e., switching takes place. Finally, depending on the fuse metal, if the energy density in the vapor reaches more than 10^{10} J/m^3, overheating of the vapor occurs, leading to formation of plasma. Formation of the plasma sets the limit on the extent of current interruption by the fuse.

The detailed physical processes that determine the behavior of a fuse involve the interaction of the fuse with its surroundings through thermodynamic and magnetohydrodynamic processes, or through atomic processes which include ionization and charged particle acceleration (arising from the large electric fields due to applied potential, or due to the potential induced by the interruption of current by the switch). These processes can be usually manipulated, in ways discussed in Chapter 8, to achieve the desired switch performance, for example, by a choice of appropriate fuse materials, the medium surrounding the fuse, and even circuit configurations.

Similar physical phenomena govern the behavior of spark gap switches. Triggered and untriggered breakdown between the electrodes, used as a closing switch, in vacuum and low pressure gases, in high pressure gases, in liquids, and in solids discussed in Chapters 4, 3, 6, and 5, respectively, is also governed by transition from the initial state of matter (in this case non-conducting) to a final (conducting) plasma state. The energy density of the breakdown channel (plasma) buried in solids and liquids is about 10^{11} J/m^3. It is somewhat lower in gases and could be substantially lower in vacuum arcs of very low pressure spark gaps. Additionally, physical phenomena associated with breakdown initiation, such as streamer formation and prebreakdown currents, and in some cases plasma dynamics (associated with triggering and various electrode phenomena) are important.

Existence of the plasma state at the completion of switch closing prevents the switch from recovering immediately, because such plasmas are difficult to cool quickly and significant residual conductivities can remain for milliseconds in the inter-electrode region. Variety of methods are employed to reduce the recovery time, including blowing of gas into the arc region and/ or magnetically pushing arcs into thermal contact with cold solid materials.

Another approach, avoiding the need to move a significant mass of the material, which is inherently slow (at most, at acoustic velocities), has also been employed. It utilizes volume conduction through a medium which changes resistance to switch the current on or off. One method uses injection of ionizing radiation into neutral gases to produce a sufficient number of free electrons for conduction. Once the radiation is turned off, the resistivity recovers rapidly. Such switches, using injection of electrons from an external source to produce ionization, are discussed in Chapter 9. To prevent the transition of the gas to a plasma state, the electron and ohmic heating of the gas is kept below the energy density level of 10^6 J/m^3 as shown in Fig. 1–2. To optimize the operation of such switches and perhaps to raise their energy density limit for recovery to $\geq 10^7$ J/m^3, an understanding of the recombination processes for electrons in gases with mixtures of complex molecules is needed.

Another approach employed in switch control through volume conduction is the use of solid materials, the so-called PTC (positive temperature coefficient) materials with non-linear resistivities, discussed in Chapter 10. Materials with high intrinsic conductivity can be tailored to change their resistivity by many orders of magnitude (e.g., due to phase change in the solid material) in response to thermal, electric, or magnetic pulses. Both single crystals and polycrystalline materials as well as plastics can be used for this purpose. Because of the higher particle density of such solids, the energy per particle remains low at the higher energy density of the switching medium, compared to the gaseous switches; the switching operation of the solids retains instanaeous recovery, since both the off-state and on-state of the switching medium retain the ability to withstand high electric fields.

Figure 1–2 also includes switches operating below 10^6 J/m^3 which also employ plasmas[5]. Because their energy density is so low, these switches (thyratrons) can be controlled by biasing the third electrode (grid). Because they cannot operate at power levels of concern in this book, they are not further considered; their listing in Fig. 1–2 is for comparison purposes.

The fundamental understanding of the many physical principles enumerated in this chapter has made the development of pulse power technology to its present state possible and, within the scope of this technology, it has led to the development of high power switches which has paid off handsomely in the construction of a variety of modern pulse power systems. Switch technology will continue to play a very important role in the development of pulse power technology. Material science and plasma physics are two disciplines destined to become keys to unlocking the solutions necessary to meet the demands of pulse power technology. The direction of these sciences in regard to application to pulse power switching is already perceptible. At present, attempts to employ the specialized materials and plasmas which can be precisely controlled are directed at developing switches for more compact pulse power sources and repetitively pulsed systems (as discussed in Chapter 10, in relation to developing material technology, and in Chapters 4 and 9, in relation to the application of plasmas). The knowledge of materials and techniques for fabricating the materials suggests that it would be possible to tailor many of their properties for specific switching requirements. Recent application of plasma physics principles and techniques to solve several switching problems suggests that this may continue through increasingly sophisticated methods of controlling various types of plasmas.

REFERENCES

1. W. Rieder, "Circuit Breakers", *Scientific American* **224**, 76–84 (January 1971).
2. T. H. Lee, *Physics and Engineering of High Power Switching Devices,* MIT Press (1975).
3. V. S. Komelkov, "Technology of Large Impulse Currents and Magnetic Fields", Foreign

Technology Division, Wright Peterson AFB, Ohio, Report FTD-MT-24-992-71. The original source volume (in Russian) is *Tekhnika Bolshilkh Impulsnylkh Tokov i Magnitnykh Poley,* Atomizdat, Moscow (1970). See also Ch. 4 references.

4. E. K. Inall, ed., "High Power High Energy Pulse Production and Application," *The Proceedings of an Australian–U.S. Seminar on Energy Storage, Compression and Switching,* ANU Press, Canberra (1977).

5. F. B. A. Frungel, *High Speed Pulse Technology,* Vol. 1, Academic Press, New York (1965).

6. T. R. Burkes, J. P. Craig, M. O. Hagler, M. Kristiansen, and W. L. Portnoy, "A Review of High Power Switch Technology," *IEEE Trans Electron Devices* **ED-26,** 1401–1411 (1979). This review describes principles of operation of many of the switches used for single pulse and repetitive operation and provides their operating ranges. It contains extensive references to more detailed discussion of switches and pulse power systems.

7. R. A. Gross, *Fusion Energy,* Chapter 9, John Wiley & Sons, New York (1984).

8. M. Buttram, "High Average Power Rep Rate Systems", *Digest of Technical Papers of the Fourth IEEE Pulsed Power Conference,* IEEE Cat. No. 83CH1908-3, (1983), pp. 361–367.

9. High power sources and their applications are described, for example, in a series of proceedings of international conferences dealing with the production of very intense charged particle beams. The most recent Conference is reported in *Proceedings of the Fifth International Conference on High Power Particle Beams* (San Francisco), R. Briggs and A. J. Toepfer, eds., LLNL CONF-830911.

10. R. B. Miller, *Intense Charged Particle Beams,* Plenum, New York (1982).

11. Methods for, and applications of the electromagnetic propulsion of mass are updated in the recent proceedings of Conferences on Electromagnetic Launch Technology, published in *IEEE Transactions on Magnetics,* **MAG-18** (1) and **MAG-20** (2) (1982 and 1984, respectively).

12. *Albuquerque Journal,* photographed by Brian Walski, sect. B, p. 1, Albuquerque, NM, 25 May 1985.

13. J. Bernesderfer, R. L. Druce, B. Frantz, A. H. Guenther, M. Kristiansen, J. P. O'Loughlin, and W. K. Pendelton, *Pulsed Power Bibliography,* Vols. I and II, Air Force Weapons Laboratory, Kirtland AFB, NM, Report AFWL-TR-83-74 (1983).

14. H. Knoepfel, *Pulsed High Magnetic Fields,* North-Holland, Amsterdam and London (1970) (same as Ref. 1, Ch. 2).

15. MLI Capacitors, Model Nos. 32317, 32318, 32383, Maxwell Laboratories, 8888 Balboa Ave., San Diego, CA 92123.

16. A. E. Robson, R. E. Lanham, W. H. Lupton, T. J. O'Connell, P. J. Turchi, and W. L. Warnick, "An Inductive Energy Storage System Based on a Self-Excited Homopolar Generator," *Proceedings of the Sixth Symp. on Engineering Problems of Fusion Research,* IEEE Cat. No. 75CH1097-5-NPS, p. 298 (1975) (same as Ref. 9, Ch. 2).

2
Switches as Components of High Power Circuits

BASIC PRINCIPLES

In pulse power systems there are two basic ways to store electrical energy, namely, in electric fields and in magnetic fields. To release the energy to the load, closing and opening switches are required to extract the energy from the electric (capacitive) and magnetic (inductive) energy storage. Often, this must be done repetitively, at high frequency. Although energy can be stored at greater density using magnetic rather than electric fields, the latter are employed much more commonly because energy extraction is performed by closing switches which are essentially state-of-the-art devices for a very wide range of performance parameters. The opening switches needed to extract the energy from magnetic fields are less developed for high power applications.

To develop an appreciation of the intimate relation between the elements comprising various high power circuits and the switches, which control the transport of the energy from energy storing elements to the load, such transport is described with network terminology and the terminology of electromagnetic power flow. The latter is needed because generation of high power involves use of elements with dimensions which are sufficiently large to make wave transit time effects important, so that lumped circuit description of the power flow becomes inadequate.

The major factors which constitute the basis for development of high power switches are the efficiency of energy delivery from the pulser to the load, shaping of the output pulses, and controlling the timing. The effects on efficiency arise, for example, from losses in the switch itself. The energy transferred to the load usually must be delivered as a pulse of given rise-time and duration. In addition, the time of delivery of the pulse frequently must be controlled accurately. Also, because several switches may be used in one network, for example to reduce the power dissipated per switch to a manageable level, close synchronization of switching is required.

Lumped Circuits

The basic circuits employing closing and opening switches are shown in Fig. 2-1. To store energy electrostatically, a capacitor is employed together with a power supply which charges it to voltage V. The energy stored in capacitor C is

$$W_c = \tfrac{1}{2} CV^2 \qquad (2\text{-}1)$$

Commercial capacitors are one form of a container for storing energy in the electric field of the solid dielectric. Energy density varies from 10 joules per kilolgram (J/kg) to more then 200 J/kg. Such capacitors can be charged slowly, to their rated voltage, maintaining the charge with only small losses over long periods of time. Generally, a low current power supply (i.e., high resistance source), is sufficient for charging such storage systems. Liquid dielectrics are used to store energy for shorter periods of time. Their charging time may be restricted to a relatively short period, requiring higher changing current, in order to maintain higher dielectric strength or to reduce conduction losses.

Once the charge is complete, the role of the closing switch shown in Fig. 2-1 is to transfer the energy to the load much more rapidly. The transfer time of energy to the load, for example, to resistance, R_L, can be very short—in case of capacitors configured as transmission lines, equal to the signal transit time in the dielectric containing the electrostatic energy. The circuit in Fig. 2-1 shows the closing switch simply connecting the capacitor with the resistor. In real circuits there is always some inductance associated with the switch as well as with the leads connecting the two elements. Because such inductances reduce the current delivered to the load and extend

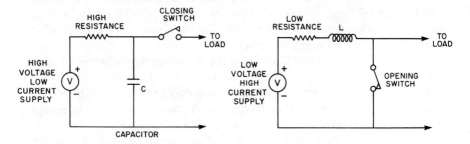

Fig. 2-1. Circuit diagrams showing the basic components for storing electric and magnetic energy and its delivery to the load.

the energy delivery time (i.e., reduce the power into the load), switch design often strives to minimize switch inductance in high power pulsed systems. (How this is done by integrating the switch with the pulser is discussed in the later section of this chapter and in other chapters.) The finite resistance of the switch leads to energy loss during transfer from the capacitor to the load. Discussion of various switches in the following chapters includes a description of switch resistance and energy loss.

To store energy in a magnetic field, a supply of current, charging an inductor shown in Fig. 2–1, is needed. In contrast to capacitive storage, the current supply operates at low voltage, but provides high current, I, through the inductor. An initially closed switch completes the circuit. The energy stored in the inductor is

$$W_0 = \tfrac{1}{2} LI_0^2 \qquad (2\text{–}2)$$

The energy density associated with inductors designed for energy storage duty is manageable at a level of the magnetic field, B, of ≤ 10 webers/m² (≤ 100 kilogauss), corresponding to $\leq 4 \times 10^7$ J/m³. Associated with the storage of energy in the form of magnetic fields is mechanical stress. In a vacuum, the force per unit volume in mks units is $F = j \times B$ where j is the local current density. This force can be used to calculate a pressure acting on the conductor. In terms of plane geometry, the pressure is $B^2/2\mu$, where permeability, μ, has a value of $4\pi/10^8$. At $B = 5T$ (50 kG), the pressure is 0.99×10^7 N/m², i.e., very close to 100 atmospheres. Thus the magnetic forces, by inducing motion of current-carrying conductors (e.g., spark channels between electrodes) play an important role in high power switches. Furthermore, consideration must also be given to the mechanical strength of switches necessary to withstand magnetic forces, to be discussed in the succeeding chapters.

Upon the opening of the initially closed switch in the circuit of Fig. 2–1, the current is forced to flow in the load circuit; in different terminology, commutation of current to the load takes place. The consequences are obvious. The voltage appearing across the load, in reaction to the current flow, also appears across the opening switch (and across the inductor and the current source). Unless the switch can withold the applied voltage, the restrike (breakdown) in the switch will result in re-initiation of current flow in the switch, preventing the full transfer of energy from the inductor to the load. The second condition for efficient energy transfer is that the switch resistance must rise to a value much larger than the load resistance, so that no significant current sharing exists between the parallel circuit branches containing the switch and the load.

A less obvious condition associated with the opening switch operation is the energy absorbed by the switch. Clearly, since the opening time is finite, i.e., the switch current is decreasing to zero in the time dictated by the processes controlling its resistance, ohmic dissipation in the switch will occur. Depending on the circuit parameters, this can be made quite small in comparison to energy stored inductively. However, if the load contains an inductor, the energy that must be dissipated in the switch could be quite large. H. Knoepfel[1], in considering various pulsed inductive storage systems, derives the energy, W_L, transferred to the load inductance, L_L, and the energy, W, remaining in the storage inductor, L, in the circuit of Fig. 2-1. The energy initially stored in L (at the time the switch is closed and its resistance $R_S = 0$) is given by Eq. 2-1, with I_0 being the current through the switch at the time just before its opening. The energy transferred to the load inductance is

$$W_L = \frac{L\,L_L}{(L + L_L)^2}\,W_0 \qquad (2\text{-}3)$$

The maximum energy transferred is 25% when $L = L_L$. The energy remaining in the inductance *after* switching is

$$W = \left(\frac{L}{L + L_L}\right)^2 W_0 \qquad (2\text{-}4)$$

so that the energy dissipated in the switch resistance is

$$W_{SW} = W_0 - W - W_L = \frac{L_L}{L + L_L}\,W_0 \qquad (2\text{-}5)$$

Thus, in switching from inductor to inductor, the opening switch leg of the circuit must dissipate 50% or more of the stored energy. Sometimes, in order to alleviate the resulting heating of the switch, a parallel resistor of sufficient mass to handle part of the switching energy is added[2]. It should be remembered, however, that as load inductance becomes small relative to the stored inductance and the energy instead is transferred to a resistive load, the energy dissipated in the switch could be quite small.

Electromagnetic Power Flow

The electromagnetic power flow description of the pulsed power sources has been discussed by P. Turchi in Ref. 3. The stored electric and magnetic energy is given by

$$W_e = \tfrac{1}{2}\,\overline{D} \cdot \overline{E} \qquad\qquad\qquad (2\text{-}6a)$$

$$W_m = \tfrac{1}{2}\,\overline{H} \cdot \overline{B} \qquad\qquad\qquad (2\text{-}6b)$$

respectively, where vectors $\overline{E}$ and $\overline{B}$ are the electric and magnetic fields. The related quantities are $\overline{D} = \epsilon\overline{E}$ and $\overline{H} = \overline{B}/\mu$, the electric displacement (ϵ denoting the dielectric constant) and magnetic intensity. For isotropic media, the Poynting theorem,

$$\frac{d}{dt}\left(\frac{\epsilon E^2}{2} + \frac{B^2}{2\mu} \right) = -\,\nabla \cdot \overline{S} - \overline{E} \cdot \overline{j} \qquad (2\text{-}7)$$

expresses the fact that the electric and magnetic energy in a volume will decrease due to net flow of the electromagnetic radiation ($-\,\overline{\nabla \cdot S}$) through the surface of the volume and to ohmic losses within the volume ($-\,\overline{E \cdot j}$).

In practical applications, either the E^2 or B^2 term dominates in the above equation. Parallel plates with an electric field applied across them (i.e., neglecting the magnetic term) are an example of a simple capacitor. Discharging such plates by closing the connecting switch sets up a wave of an opposing field. The plates present the characteristic impedance of a transmission line to the wave. If the discharged energy is absorbed in a load and switch with a combined impedance equal to that presented by the plates, then in one round-trip (transit) time, all the energy is removed from the plates[4].

These plates, when they are connected at both ends (i.e., short-circuited), cannot support any electric field. However, a current source can be used to initiate a circulation of the current, filling the volume between the plates with a magnetic field (with the electric field term in Eq. 2-7 being negligible). Effectively, such a loop now is a magnetic storage device. Breaking the connection between plates generates an electric field with an amplitude given by the circulating current times the impedance consisting of the load connected across the plates in parallel with the impedance of the plates. This field will propagate to the other end of the plates opposing the existing flow of current. Again, for a load resistance matched with the plate impedance, the energy will be fully absorbed in one round-trip transit time, resulting in cessation of current flow.

Both of the discussed cases are idealized. Inability to match because of varying switch resistance, additional inductance, and load mismatches leads to wave reflections, which spread out in time the delivery of energy to the load[4]. As the time of energy delivery increases, the system can be described more simply by the lumped circuit equations. This description becomes accurate as the energy delivery time becomes much greater than the transit time associated with the largest dimension of the storage system.

HIGH POWER SWITCH TECHNOLOGY

The switches for high power applications often must be designed as integral parts of an entire pulse power system. The system's electrical characteristics in many circumstances are influenced by geometric relationship of the switch electrodes to other parts of the circuit. Mechanical forces associated with the switches, arising from the magnetic currents in the switches or due to heating by switch arcs, can be very substantial. The switches' electrical design must make compromises with their mechanical requirements. Erosion of electrodes, contamination of insulating surfaces, and formation of undesirable debris must be factored into the design in order to prevent degradation of switching performance where long switching life is desired.

Closing Switches

Closing switches are described in Chapters 3, 4, 5, and 6. Their main role in very high power systems is to hold off the voltage during its build-up and to close, i.e., allow the current to flow at designated level and time. A switch performing this role is shown in the circuit at left in Fig. 2-1. Practical switches have inductance, resistance, and possibly significant stray capacity. They often need triggering capability. The relevance of these characteristics depends on the remainder of the circuit and its function. Sometimes, at least, some of these quantities can be neglected.

An example of an integrally-designed switch for a pulser system with an output greater than 10^{12} W is the output switch developed by J. D. Shipman, Jr., and used in the Naval Research Laboratory's GAMBLE II generator[5], shown in Fig. 2-2. (The principles of operation of this switch, capable of transporting power to the load at a 10^{12} W level, are discussed in detail in Chapter 6.) The GAMBLE II generator uses a large diameter single low impedance pulse line to produce an output with a risetime of ~25 ns and duration of about 70 ns. The voltage across the matched load reaches more then 1 MV and the current through the load is 1 MA. To achieve such performance in a sufficiently compact system, water is used as an insulating dielectric. To avoid electrical and mechanical problems usually associated with the insulator interfaces, which would occur if the medium between switch electrodes were a gas (or vacuum), requiring walls to separate such a medium from the surrounding water, the closing switches were designed using water as the dielectric as described in Chapter 6. Furthermore, as discussed in the design manual for the GAMBLE II generator[6],

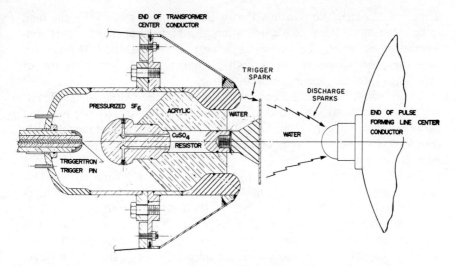

Fig. 2-2. Triggered high-current water breakdown switch illustrating careful design required for low switch inductance and low jitter operation at voltages of several megavolts.

even though the output impedance of the generator is 1.5 Ω, the output switch was located at the output of an intermediate 5.8 Ω pulse line section. The power stored on this line was switched into matched coaxial transmission line of variable impedance (decreasing from 5.8 Ω, away from the switch), connecting it with the load, with a match at 1.5 Ω. Because the energy stored on the pulse line is

$$W_{\mathrm{PL}} = \frac{V^2\, t_{\mathrm{P}}}{4Z} \tag{2-8}$$

where t_{P} is the pulse duration, Z is its impedance and V is the applied voltage, with W and t_{P} specified by the applications intended for the pulse line, it follows that

$$V \propto \sqrt{Z} \tag{2-9}$$

Since the switch inductance, L, is proportional to the separation of the electrodes, $V \propto L$, so that the risetime of the switch, τ, is,

$$\tau = \frac{L}{Z} \propto \frac{1}{\sqrt{Z}} \tag{2-10}$$

Thus, to minimize the risetime, the switch is located away from the load in order to allow the switch's operation at a higher impedance. This configuration, however, also implies in the case of the GAMBLE II generator that the switch operates at a higher (input) voltage (by a factor of $\sqrt{Z_{in}/Z_{out}}$) than the designed output voltage.

Additionally, because the energy dissipated in the switch is significant, the pulse line transformer also provides a very effective acoustic isolation between the switch explosion, caused by the arcs in water, and the relatively fragile electron beam vacuum diode, normally used as a load for the GAMBLE II generator.

The circuit in Fig. 2–3 represents a different use of switch arrays. This circuit, called the Marx voltage multiplier circuit, is used to obtain voltages as high as 10 MV. These arrays, such as the four stage unit shown in Fig. 2–3, use low voltage capacitors, (or pulse lines), C, which are charged in parallel to voltage V_0 and discharged in series. The individual charging voltages of n capacitors are added up at the output to obtain the total voltage: $V = nV_0$ (if there are no losses). The specific role of the switches in such arrays is to hold-off the voltage during charge (at low rate of current flow through the resistor R_0 and through choke inductances, L). Once the capacitors are charged, the switches $S_1, S_2, \ldots$, are closed, resulting in the generation of a high voltage output across the load. Typically, because the discharge time is many orders of magnitude less than the charging time, the Marx generator output current is much higher, requiring the switches nearer to the output end to be able to withstand high currents. The switches are closed by triggering several of them in the initial stages. The trigger

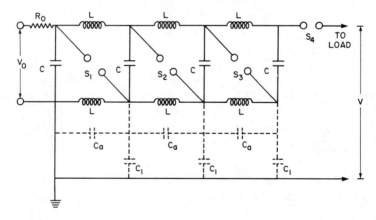

Fig. 2–3. Circuit diagram for the Marx generator, showing the main energy storage elements, C, charged in parallel, charging inductors, L, together with the stray elements C_1 and C_a. Closing switches $S_1, \ldots$, are used for discharging the capacitors by connecting them in series.

level and the number of switches which must be triggered simultaneously to produce maximum output depends sensitively on the stray capacities C_a . . . and C_1. . . . The output switch S_4 serves to isolate the load from the current charging the capacitors. Discussion of such circuits can be found in Ref. 4.

The circuits such as those in Fig. 2-3 and high current circuits using parallel capacitors, such as the PHAROS bank[7] shown in Fig. 2-4, can be thought of as utilizing the switches to distribute the energy required for the output pulse to a level where its density is sufficiently low for safe and reliable handling. In the PHAROS bank, which ties a large number of capacitors in parallel to obtain large currents, the switches are connected in parallel, in contrast to series connection in the Marx circuit. The ability to tailor the energy density or power density consistent with a state of development of the component technology is an exceedingly important factor in very high power pulse power systems.

Not all closing switches are used to switch at the maximum of the applied voltage. The so called "crowbar" switches are turned "on," i.e., they allow current to flow at negligibly low voltage across the switch. These often are used in conjunction with diverting current from the capacitors (at the time of peak current and minimum voltage). Their use, design, and operation in high current capacitive storage systems are discussed in Chapter 5.

Opening Switches

The integration of an opening switch into a practical pulse power system presents no special difficulties at low stored energy and low input power. As the energy to be switched increases, the energy dissipation in the switch often becomes a dominant problem in designing the inductive circuits. Distribution of explosive pressures over manageable volume by paralleling many opening switches is employed as one approach. Some damage to the switch hardware, such as holders and current contacts, is tolerated in many circumstances. The methods that have been developed are discussed most thoroughly in Chapter 8, in conjunction with the use of exploding wires, or fuses, as opening switches. The analysis of switches as part of a circuit and of the physical characteristics of switches appears in Chapter 9, where one specific type combines both of these factors, serving as an example to show how these factors enter into a full description of switch behavior.

Two large pulse generators utilizing well developed opening switch techniques are the Air Force Weapons Laboratory SHIVA[8] and Naval Research Laboratory TRIDENT II[2]. These two generators are representative prototypes of quite different approaches to using inductive storage at a multi-megajoule energy level. The SHIVA pulser uses capacitors to provide

Fig. 2-4. A 2 MJ capacitor bank, using parallel switches (shown emitting light during operation) to reduce the inductance which limits the current. Reproduced from Ref. 7.

multi-megampere current to a small inductor in about 3 microseconds. The interruption of the current in the inductor by a large parallel array of fuses effectively steepens the output pulse, enabling the generator to deliver several terawatt pulses to plasma loads[8]. The TRIDENT II pulser, shown in Fig. 2–5 uses a homopolar generator to energize an inductor with a maximum current of about 70 kA. The current peak is achieved, depending on the current level, in a time of about 1 second. Thus, the opening switch in the inductor circuit must carry the current for a time many orders of magnitude greater than the time in the SHIVA pulser. Because the switch resistance must be very low during conduction, in order to obtain efficient excitation of the homopolar[9], explosively-driven switches, discussed in Chapter 7, are used employing the staging technique. This technique of commutation from one opening switch, with good long-conduction capability and poor (long) opening time, to the next switch, with short conduction time capability and better opening performance, has been developed in the USSR[10] and in Germany[11]. Figure 2–6 shows such staging in a circuit

Fig. 2–5. Naval Research Laboratory compact self-excited homopolar generator and inductive energy storage system. Reproduced from Ref. 2.

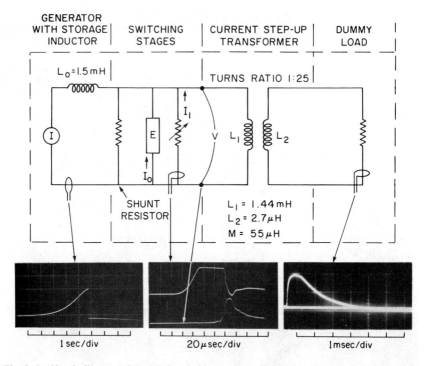

Fig. 2-6. Circuit diagram of the HPG switching and transformation elements for high voltage and high current generation. Also shown are current and voltage traces.[2] The first trace shows the excitation current (36 kA). Next are shown the fuse current (I_1 = 35 kA) and voltage (V = 200 kV) developed by the fuse. The last trace is the secondary current through a 10^{-3} Ω load with 420 kA peak amplitude.

associated with the system shown in Fig. 2-5. Employing the multi-stage switching techniques, it has been shown that long-duration current sources, of which the homopolar is a typical example, could be used to power an inductive storage device for multi-terawatt output[12] similar to that provided by capacitive energy storage.

The principles of operation of staging several opening switches in succession and typical time scales are given in Fig. 2-7. Such staging is commonly used because there is no one opening switch which can satisfy the usual requirement of the inductive storage systems requiring long conduction time and short opening time. It involves initial circulation of current in the inductor through the first opening switch (e.g., "circuit breaker," shown in Fig. 2-7). Its low resistance is required in order to minimize ohmic losses during circulation (i.e., during "charging" of the inductor). As peak current is reached, the breaker is made to open, commutating the current to

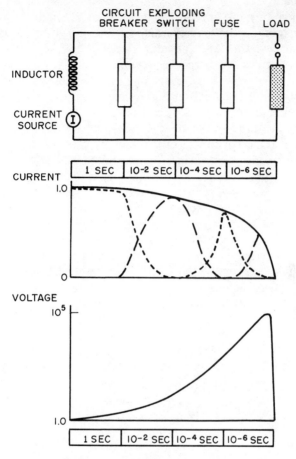

Fig. 2-7. Idealized schematic of a power multiplication in inductive storage circuits. Also shown are the current and voltage waveforms demonstrating sequential transfer of current from one switching stage to the next and, ultimately, to the load.

the next stage (possibly through a closing switch, not shown in the figure, used to isolate the current from the second stage during charging). The second stage, an explosive switch, or perhaps directly the fuse, conducts the current for much shorter time, e.g., shorter by a factor of 10^{-2} for the exploding switch or even a factor of 10^{-4} for the fuse. When full current is reached, the second stage opens (due, for example, to vaporization of a fuse, as discussed in Chapter 8), transferring the energy to the load. If there are no losses, the current will reach the same peak (the relative amplitude of 1 in Fig. 2-7) in each stage. Due to losses, the total current (solid curve) is somewhat reduced before switching to the load occurs. The current in the inductor is constant, or reduced by switch losses, until it begins to decay due to dissipation in the load. Because of increasingly faster interruption of the current by each stage, the (inductively generated) voltage across the inductor (and across the current source) increases. This increase would be proportional to the switch stage opening time. Again, due to losses some reduction occurs in the final output voltage appearing across the output closing switch and the load.

As in the case of the Marx circuit, employing parallel charging and series discharging of the capacitors to obtain multiplication of the charging voltage by the number of series stages, similar multiplication of current is also possible. This approach, shown in Fig. 2–8, is aimed at generating high current output using low current sources. It is based on series charging of the inductor, with parallel discharge, so that the output current is a multiple of the number of the inductor stages[13]. This is sometimes referred to as an inverse Marx circuit or XRAM circuit. The final output opening switch can be staged (as represented by the two variable resistors) to provide high output voltage, as shown in the figure. As a first large scale application of the current multiplication, I. A. Ivanov et al.[14] have designed a 20 MJ storage system for thermonuclear fusion studies requiring large currents. To achieve such currents, they employ an XRAM circuit with current multiplication of 30.

Finally, there is one other type of device, consisting of a section which is accelerated by the magnetic field, which in turn functions as a current transfer device. This device is thus categorized in a broad sense as an opening switch. This device is a conductor (metal or plasma) which is allowed to move under the forces arising from the current used in charging of the inductor. For example, a conducting plug, allowing a current to flow be-

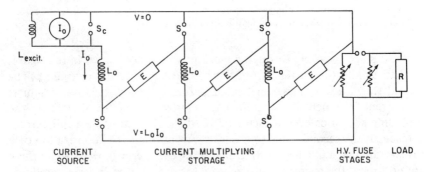

Fig. 2–8. Current-multiplying circuit. The current source, a self-excited homopolar generator (using inductor $L_{excit.}$), charges a series of inductors, L_0, with current I_0. After charging, crowbar switch S_c is closed to take the current source out of the circuit; the opening of switches E and the simultaneous closing of switches S connect the inductors to the load.

tween the inner and outer cylinders of a coaxial system (or between two parallel plates), can be made to move along the axis increasing the inductance of the coaxial (parallel plate) inductor[15]. The rate of change of inductance is analogous to insertion of the resistance in series with the inductor, leading to an increase in the voltage in such a circuit. If the plug is driven out the end, a physical break in the conductor occurs leading to the appearance of inductive voltage. Such interruption at high power levels is very complex due to plasma formation at the moving contacts and to generation of arc plasma when the break in the contacts occurs. The status of such switching is discussed in Chapter 9.

Repetitively Operated Switches

A different class of switches also employed in pulse power systems are the repetitively operated switches. For high power operation, the use of such switches usually is limited to production of pulse trains with a relatively small number of high peak power pulses generated in succession. The performance during "on" and "off" periods of such switches allows the energy storage to release energy in increments representing only a fraction of the total energy. The circuit for such energy storage would not be different from single pulse generators. However, repetitively operated switches, capable of operating at high repetition frequency, generally differ significantly from single shot switches. The former class of switches must survive each closing-opening (or its inverse) cycle without change in their switching characteristics. This implies that it must return to conditions that are nearly the same as those before switching. Methods to achieve this, at least for a small number of switching operations, are discussed in Chapter 9.

CONCLUSION

The interaction between circuit parameters and switching characteristics plays a central role in solving high power switching problems. As switch technology progresses, it becomes practical to employ energy storage circuits with broader operating ranges as well as to improve the efficiency of energy transport from the storage element to the load. In larger systems, switch costs and circuit complexity (and therefore reliability) also are being reduced with advances in the technology.

REFERENCES

1. H. Knoepfel, *Pulsed High Magnetic Fields,* North Holland, Amsterdam and London, 1970.
2. R. D. Ford, D. Jenkins, W. H. Lupton, and I. M. Vitkovitsky, "Pulsed High-Voltage and High-Current Output from Homopolar Energy Storage System," *Rev. Sci. Inst.,* **52,** 694 (1981).

3. P. J. Turchi, "Inductive Energy Systems," *Pulsed Power Lecture Series No. 13, Plasma and Switching Laboratory Report,* Texas Tech Univ., Lubbock, TX (undated).

4. Behavior of waves due to switching of transmission lines into mismatched, inductive, and capacitive loads as well as use of transmission lines for pulse shaping is discussed by I. A. D. Lewis and F. H. Wells in *Millimicrosecond Techniques,* Pergamon Press, New York (1959). A review of transmission line characteristics and applications for those geometries more commonly used in pulse power systems is reported by D. L. Smith, "A Transmission Line Survey," *Pulsed Power Lecture Series No. 12, Plasma and Switching Laboratory Report,* Texas Tech Univ., Lubbock, TX (undated).

5. The pulse generator using the switch shown in Fig. 2–2 is described by L. S. Levine and I. M. Vitkovitsky, *IEEE Trans. on Nucl. Sci.* **NS-18,** 255–264 (1971). The switch design was originally described in the note by J. D. Shipman, "Triggered Water Switch Development for GAMBLE II," Naval Research Laboratory, Washington, DC, unpublished.

6. John D. Shipman, "Final Electrical Design Report of the GAMBLE II Pulse Generator," NRL Memo Report 2212, Naval Research Laboratory, Washington, DC (1971).

7. A. C. Kolb, H. R. Grien, W. H. Lupton, D. T. Phillips, S. A. Ramsden, E. A. McLean, W. R. Faust, and M. Swartz, "A High Energy Magnetic Compression Experiment," *Nucl. Fusion,* Suppl.—Part 2, p. 553 (1962).

8. R. E. Reinovsky, W. L. Baker, Y. G. Chen, J. Holmes, and E. A. Lopez, "Shiva Star Inductive Pulse Compression System," *Digest of Technical Papers of Fourth IEEE Pulse Power Conference,* IEEE Cat. No. 83CH1908-3 p. 197 (1983).

9. A. E. Robson, R. E. Lanham, W. H. Lupton, T. J. O'Connell, P. J. Turchi, and W. L. Warnick, "An Inductive Energy Storage System Based on a Self-Excited Homopolar Generator," *Proceedings of the Sixth Symposium on Engineering Problems of Fusin Research,* IEEE Cat. No. 75CH1097-5-NPS, p. 298 (1975).

10. V. A. Zheltov, A. V. Ivlev, A. S. Kibardin, A. V. Komin, V. G. Kuchinskii, and Y. A. Morozov, *All-Union Conference on Engineering Problems of Thermonuclear Reactors,* Leningrad, NIIEFA Preprint K-0297 (1976).

11. J. Salge, U. Braunsberger, and U. Schwarz, "Circuit Breaker for Ohmic Heating Systems," *Proceedings of the Sixth Symposium on Engineering Problems of Fusion Research,* IEEE Cat. No. 75CH1097-5-NPS, pp. 643–646 (1975). Also, by the same authors, "Circuit Breaking by Exploding Wires in Magnetic Energy Storage Systems," in *Energy Storage, Compression and Switching,* pp. 477–480, Plenum, New York (1976).

12. R. D. Ford, D. J. Jenkins, W. H. Lupton, and I. M. Vitkovitsky, "Multi-megajoule Inductive Storage for Particle Beam Production and Plasma Implosions," *Proceedings of the Fourth International Topical Conference on High-Power Electron and Ion-Beam Research and Technology* (Palaiseau, France), Commisariat a l' Energie Atomic–D.A.M. and Ecole Polytechnique, p. 743 (1981).

13. Kernforschungsanlage Jülich GmbH, German Federal Republic Patent 1488941 (1972).

14. I. A. Ivanov, A. P. Lototskii, N. P. Pugachev, and V. A. Trukhin, "High Power Three-Stage Switch for an Electrical Discharger with Inductive Energy-Storing Element," *Instrum. and Exp. Techniques* **25** (1), 895–897 (1982).

15. D. R. Kania, L. A. Jones, E. L. Zimmerman, L. R. Vesser, and R. J. Trainor, "Experimental Investigation of a Magnetic Gate as a Multimegampere, Vacuum Opening Switch," *Appl. Phys. Letter* **44,** 741 (1984).

3
Gas Spark Gap Switches

INTRODUCTION

In this chapter, gas spark gaps are considered. Spark gaps are switching devices most widely used in production of high power pulses. Their unique feature is the ability to switch low currents as well as currents in the multi-megampere regime. In this chapter, spark gap switches where the insulating medium is a gas at atmospheric or higher pressure are considered. Spark gaps using low pressure gas are discussed in Chapter 4. Self-break and triggered spark gaps are used in different modes, depending on the function to be performed by such switches. There are two major modes of spark gap operation: the use of the switch to hold-off the applied potential until breakdown (or triggered breakdown) of the gap between the switch electrodes occurs, resulting in current discharge; and, less frequently used, the crowbarring mode, in which the gap is made sufficiently conducting by some means (e.g., by injecting a burst of plasma to ionize the gas in the gap) so that the current flows through the switch at practically zero voltage across the gap.

Simple spark gap switches consist of two electrodes separated by a gap. The electrodes are supported by a structure made of insulating materials, or materials forming part of the connections to the circuit. Such gaps can be stacked in series for, as an example, higher voltage hold-off. The supporting structure may serve an additional function of containing a gas, perhaps at other than ambient pressure, that provides a desired switching characteristic. Switching without triggering is performed by applying an increased electric field across the electrodes until the breakdown occurs. The discharge following the breakdown is a narrow plasma channel (or multiple channels) characterized by a resistance, R_s, which is achieved in time, τ_R, after channel initiation. Because of the channel diameter, channel inductance, L_s, becomes significant in fast-discharge capacitor banks and in discharging of low impedance pulse lines.

The channel conductivity can remain at sufficiently high level, even after the current ceases to flow through the channel. The duration of high conductivity level, i.e., the recovery period, is an important factor in selecting the type of switching device for circuits used in conjunction with the repetitive pulse operation. The spark gap switches are low loss, self-healing switches which can be operated continuously at duty cycle determined by the decay of channel conductivity (which for simple gaps carrying peak currents of kiloampere level is in the range of 10 ms). On the other hand, the number of reproducible discharges may be limited by the life of the switch determined by the erosion of electrodes, deposition on insulating supports, and contamination of the insulating gas, as well as overheating.

Extensive literature dealing with spark gap switches exists. A large fraction of the literature consists of unpublished reports and of low circulation conference proceedings, notably those of bi-annual Modulator Conferences[1] and Pulse Power Conferences[2]. Review of spark gap switch designs and performance, as well as some discussion of scaling laws applicable in high power regime are discussed in Refs. 3, 4, and 5, published recently, as well as in older sources. Several patents have been granted featuring specific design objectives[6, 7, 8, 9]; one of them, for example, uses magnetic blowout[8].

The design objectives and goals, such as the capability to carry high current or large amount of discharge, to withstand very high voltage, and to provide very reliable and long-life operation and precise triggering are reflected in the choice of specific design characteristics. High current and large amount of charge can be handled by forcing the discharge to move along the electrode surface during the discharge[7], while reducing electrode erosion by spreading the arc heat over a large electrode area. Such a switch has been employed in a 2 MJ Marx generator[10] at high current levels near 1 MA. This technique is also used in switches with gas pressures lower than atmospheric; these switches are described more fully in Chapter 4. A somewhat different approach to high current switching is taken in a design where the arc is blown away from an insulating wall[8], preventing accumulation of debris on the surface and effectively extending the switch's lifetime very considerably[11]. For large operating ranges, a device consisting of a number of series gaps, each connected to a trigger pulse, has been developed[9]. The nomenclature associated with various types of switches and methods for achieving specialized objectives such as low electromagnetic noise or low prepulse are discussed later in the chapter.

DESIGN CRITERIA

The design of gas spark gaps for switching at high power levels consists of accommodating into one package the following elements: gas dielectric, electrodes with their support structure, triggering elements (if pulser design consideration require it), and other special features such as a means for flushing contaminated gas. These elements are combined by an appropriate design into a switching device that answers a given requirement of the pulser system. Often, these requirements, discussed to some extent in Chapters 1 and 2, include a specified switching voltage, current carrying capability, total charge transfer, and such practical considerations as reliability, lifetime, cost, and fabrication constraints. In some circumstances, it is more practical to build switches by combining either parallel or series gaps using separate modules or integrated packages. Such approaches can be found in Refs. 11 and 9. Additionally, selection of a triggering scheme, if such is to be employed, requires careful consideration of the allowable delay period between trigger pulse and switch initiation and the jitter in the delay. However, the starting point for designing the switch is the breakdown property of the gas which provides the insulation in the gap.

Gas Breakdown

Gas breakdown can be defined as the transition from a very high resistivity state of about 10^{14} Ω-m to a high conductivity state[12], caused by ionization due to the applied electric field across some region of gas. The ionization growth is due to the appearance of an initial free electron being accelerated in the electric field. The rate of ionization and, consequently, the resulting type of breakdown depends on which of the several processes dominates. As discussed in Chapter 1, gases can be made conducting over large volumes, or as narrow channels, i.e., spark discharges. Since spark gaps operate as switches with breakdown being of the latter type, at least when limited to those gaps involving gases at atmospheric or higher pressures, many of the possible mechanisms leading to conduction are not important. The dominant processes, described, for example, by F. L. Jones[13], may be an avalanche (exponential growth of free electrons) followed by a self-sustaining (Townsend) discharge. The avalanche cannot be sustained below critical fields (of a few percent above breakdown voltage, V_{br}) without secondary processes. These processes are ionization by positive ions (β-process), photo-ionization (becoming negligible at lower pressures), secondary emission at the cathode due to incidence of ions (γ) or excited atoms (ϵ), and the photoelectric effect (δ). The rates α through ϵ predict[12] the current in the gap, Townsend's criterion for breakdown, and ultimately the functional relationship between V_{br} and the electrode separation, d. The V_{br} dependence on d was initially observed more than a century ago[13] and developed into similarity relationship involving gas density, n, and electrode separation, d:

$$V_{br} = f(nd)_{br}, \qquad (3\text{-}1)$$

known as Paschen's law. Here f represents a function of the product of n and d at which breakdown occurs.

Technologically, the Paschen regime can be divided into that for low values of nd and that for high values of nd. For low values, i.e., for pressures roughly less than 1 torr, the processes involving collisions with electrodes play an important role; this criteria is used in Chapter 4 to distinguish the vacuum and low pressure regimes. As the pressure increases, the collision processes in the gas become more important. The spark gaps, considered in this chapter, have their performance characterized mainly by the gas properties in contrast with those discussed in the next chapter. The electrodes play only a secondary role in terms of their effects on prebreakdown distribution of field and material erosion which, for example, affects switching reproducibility. The effect of electrode geometry on the electric fields in the gap is discussed under "Electrodes and Support Structures" in this chapter.

Because of the time-dependence of the rates associated with the processes involved in the development of breakdown, the rapidly rising electric fields and slow or dc fields lead to different breakdown values. In high power applications, the applied electric field is often of relatively short duration, restricting the range of phenomena (for example, eliminating corona effects), thus simplifying the relations that can be used in the design of the spark gap switch. However, empirical scaling for dc breakdown also has been developed. J. C. Martin[14] has developed from empirical data such scaling for up to hundreds of kilovolts. He has provided such scaling for air in uniform and weakly non-uniform electric fields. For uniform fields, the breakdown can be expressed in the form

$$E = 24.6 + 6.7 \sqrt{P/d} \quad (kV/cm). \qquad (3\text{-}2)$$

In this expression, P is air pressure in atmospheres and d is electrode separation in cm. This relation shows that for gaps with separation of the order of 1 cm operating at 1 atmosphere, the breakdown field is about 30 kV/cm. As the separation decreases, the breakdown field can increase significantly.

For air, the non-linear dependence of breakdown on pressure becomes significant at small gap separation. J. C. Martin notes that when good insulating gases such as CF_2Cl_2 (Freon) or SF_6 are used, the pressure dependence is very considerable[14]. For these gases, the saturation in breakdown strength is only one of the practical limits on the maximum fields that can be maintained across the switch. Liquification of the gas occurs at relatively low levels, about 4.9 atm at room temperature for Freon and up to 24 atm for SF$_6$. Higher pressures can be achieved for mixtures of these gases with air or nitrogen[4].

Following further J. C. Martin's analysis of gas breakdown data, the scaling for very divergent fields associated with points or edges is roughly[14]

$$E^{\pm} \, (dt)^{1/6} = k \pm p^n \qquad (3\text{-}3)$$

for gaps greater than about 10 cm. In this expression, $E^{\pm}$ is the mean electric field (of positive or negative polarity at the sharp electrode) in kV/cm, d is the gap length in cm, t is the applied pulse time in microseconds and p is the gas pressure in atmospheres. For a range of 1 to 5 atm, the values of $k^{\pm}$ and n are given in Table 3-1.

In applications where pulsed voltages are applied across the gap, in uniform or non-uniform fields, the breakdown voltage can rise significantly if the applied pulses are sufficiently short as already mentioned in this chapter. Interestingly, J. C. Martin[14] also points out that time dependency for air disappears for times longer than 1 μs for negative pulses and for several hundred μs for positive polarity. Thus, for air at 1 atm, a point plane gap will require about 1.5 MV to close across 100 cm in 0.1 μs. Two major reasons for this are the statistical delay induced by the time of appearance of the initiating electron to start the Townsend avalanche and the streamer formation time[14]. The statistical delay may add a few percent increase in breakdown over dc voltage. The latter effect, the streamer formation, is discussed below in connection with the discharge channel formation. For electrodes, commonly used in pulse power applications, i.e., with reasonable areas and substantial roughness associated with multiple discharges, the effect of streamer formation delay in air can lead to a breakdown field increase of more than a factor of 2 for 10 ns pulses. This is evident for data of Ref. 15. For 1 μs pulse, the increase is reduced to few percent. The effect also decreases for higher pressures, being significant only for specialized applications.

Table 3-1. Coefficients for Gas Breakdown Scaling.*

	Air	Freon	SF6
k^+	22	36	44
k^-	22	60	72
n	0.6	0.4	0.4

*Reproduced from Ref. 14.

Discharge Products

Repeated discharges of gas gaps results in changed performance of the switch. The major reason for this is the chemical change occurring as a result of gas decomposition and reaction of the various discharge products with the electrodes and structural dielectric components of the switch. Appropriate designs also are employed to control the deposition of some of these products in undesirable areas, such as on the walls where large electric fields may exist. (Potential for creation of toxic substances resulting from discharge also exists. This is briefly discussed in the Appendix).

In general, brass is an excellent electrode material, since, in addition to ease of machining and good erosion rates, it is chemically compatible with such gases as air and SF_6 and their mixtures. To maintain reproducible performance, the flow of these gases is used extensively, usually flushing switches with a modest flow. To reduce further corrosion effects and other chemical reactions, the gas should be dried, especially if the switch is used for transferring large total charge (or, as often referred to, for "large coulomb" usage). Dielectrics such as Lucite and nylon are relatively inert so that the deposits on their surfaces are amenable to being cleaned. On occasion, as in the case of the GAMBLE II Marx, which uses about 100 high current gas gap switches that hold off up to 60 kV[16], a discardable Lucite liner can be employed to simplify switch maintenance. The various flow rates and types of insulators were studied at Culham Laboratory to determine their effects on prefires of MARK II and MARK III switches[17] for high current (~ 500 kA) and high charge transfer (10 C), low inductance (15 nH) applications at 60 kV. These switches are described later in this chapter in the discussion of triggering methods, as an example of integrated design and testing effort performed for a specific application in high performance compact capacitor banks requiring closely connected parallel switches. The two reports of Ref. 17 represent a significant step in the development of spark gap switches.

Other Influences on Breakdown

Additional effects which may have significant influence on the spark gap breakdown voltage are the temperature of the gas and corona.

In most spark gap applications, heating of the insulating gas is insuffi-
_cient to be of concern. In special situations such as operation in a high temperature environment or the use of a sequence of rapid discharges, it is possible to heat the gas to the point where the effect on breakdown becomes noticeable. Certainly, heating of gas in repetitive switches such as those

based on volume discharge, as discussed in Chapter 9, is a very serious concern. In volume discharge switches, the main effect appears due to the expansion of the heated gas on a time scale approximately equal to pulse-to-pulse separation time, with a resulting shift to a lower breakdown range on the Paschen curve. In this chapter, breakdown degradation of gas due to incipient ionization resulting from heating is considered. T. H. Lee[18] notes that nitrogen under static conditions behaves according to Paschen's law for gas temperature below about 1100°K. In experiments with shock-heated nitrogen[18], electric fields were applied, showing the beginning of significant breakdown degradation at about 2000°K. The transition to lowered breakdown occurs at electron density in the range of 10^9 to 10^{10} cm^{-3}.

The effects of corona on the performance of the spark gaps operating at atmospheric pressure can be quite severe in terms of lowering breakdown voltage and reproducibility as well as interfering with charging of the various circuit elements by low current power supplies. Corona is associated with enhanced electric fields due to presence of sharp points or edges. If sufficient field enhancement occurs, field emission leads to low level current which may be sustained over a long period. The current supplies the electrons or ions for the space charge near the electrode surface and radically changes the field distribution in the region of the neutral gas. The corona mechanisms are thoroughly discussed by J. M. Meek and J. D. Craggs[12] for large variety of gases, gas pressures, and applied voltages. Often, the corona streamers can be seen in subdued lighting. Because the existence of corona is the result of release of free electrons, various gases with electron-attaching molecules can reduce or eliminate corona. Relatively small amounts of gases such as O_2 or SF_6 make switch breakdown more reproducible by reducing the corona effects. Figure 3–1, reproduced from Meek and Craggs[12], illustrates typical voltage-current characteristics for negative point-to-plane coronas in air at atmospheric pressure, for gap separations, and voltages often of interest in high power applications.

Electrodes and Support Structure

The geometry of electrode surfaces, including those of trigger electrodes, along with the dielectric strength of the insulating gas used in the gap determines switch breakdown voltage. The supporting structure, although to a smaller degree, also influences the breakdown of the switch. These structures, made of dielectrics, can accumulate debris from the discharge or from an environment (e.g., dust) on various surfaces which have a component of electric field along the surface. If the surface fields exceed critical "flashover" value, a surface breakdown occurs. Because debris accumulation may

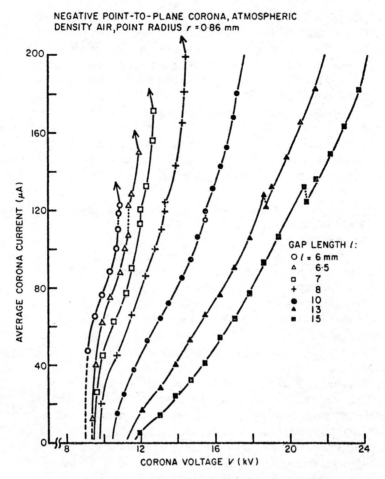

Fig. 3-1. Typical voltage-current characteristics for negative point-to-plane corona in atmosphere density air. The small discontinuity in the characteristics marks the transition from lower current Trichel pulse corona to higher current pulseless glow. Reproduced from Ref. 12, p. 351. Copyright © 1978 by John Wiley & Sons, Ltd. Reprinted by permission of John Wiley & Sons, Ltd.

collect gradually, a given switch performance would degrade slowly over a large number of discharges. Breakdown on surfaces in vacuum is described in Chapter 4.

Figure 3-2 shows an idealized spark gap mounted in an insulating cylinder. The selection of the switch design obviously depends on the intended application of the switch. Thus switches that are to be used in large numbers and in equipment where they are difficult to service, as, for example, in

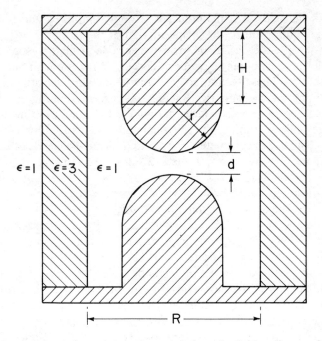

Fig. 3-2. Spark gap switch consisting of rounded electrodes (with radius r and length H), separated by a gap of length d. The electrodes are supported on plates mounted at both ends of an insulating cylinder of diameter D (with a dielectric constant $\epsilon = 3$). The cylinder contains an insulating gas with $\epsilon = 1$.

large Marx banks, must be carefully designed by taking into account the distribution of electric fields to minimize erratic breakdowns and to extend their operational lifetime. Both the electrodes and the insulating electrode support structure, as well as various trigger accommodations, determine the details of the electric field distributions. Generally, in addition to establishing the electric fields between the electrodes which determine the self-breakdown voltage of the gap and the flashover fields on the dielectric surfaces, it is also necessary to pay attention to fields formed at the so-called "triple points," i.e., in regions where the conductor, the solid insulator (with a dielectric constant greater than air), and the insulating gas meet. The "triple points" appear at both ends of the cylindrical insulator housing of the switch shown in Fig. 3-2. Improper design of the "triple point" region may lead to flashover, to breakdown through the insulator, or to a gradual degradation of the insulator by formation of carbonized dendrites. Because spark gaps are sometimes used in water (as a dielectric, with dielectric constant $\epsilon = 80$ in the construction of large capacitors[16]), the intensity of the fields at triple points is strongly amplified and this may be exceedingly important to the proper design of the switch.

To assist with the design of proper electrode-insulator geometries, potential plotting techniques have been developed for use with modern computer facilities. These techniques are now made very versatile and can be used with complex geometries, including those of floating electrodes (as may be needed in various trigger designs)[19]. Potential plotting code[19] can be used as an illustration of the electrical potential distribution in five cases, with the geometrical parameters chosen to indicate the trends of the electric field changes. For five cases, Table 3-2 summarizes the values of the maximum electric fields in a gap, $E^{(g)}$, and on the inner, $E^{(in)}$, and outer, $E^{(out)}$, surfaces of the insulating cylinder which separates the two electrodes. Also, for each case, the average field in the gap, $E^{(g)}_{av}$, is given. The variable switch dimensions shown in Fig. 3-2 are the length of the cylindrical section of the electrode, H, the electrode gap, d, and the inner diameter, R, of the insulating cylinder. The dielectric constant of the insulator is $\epsilon = 3$. The remaining dimensions of the switch can be scaled from Fig. 3-2. Inside and outside the cylinder, the dielectric constant of the insulating gas is $\epsilon = 1$.

The resulting electric fields in the electrode gap and on insulating surfaces have been determined during the calculation of the field distributions (shown in Figure 3-3, for Cases 3 and 4 listed in Table 3-2). Figure 3-3 represents computer graphing of the potential distribution for two values of insulating cylinder diameters. Case 3 shows an insulating sleeve placed directly over the electrodes. Case 4 shows the insulating sleeve diameter to be twice that of the electrodes. Maximum and average electric fields in the gap, $E^{(g)}$ and $E^{(g)}_{av}$ respectively, are seen to be approximately the same except for the case when a much longer gap (Case 5) is used. At the same time, the maximum electric fields on the inside and outside surfaces of the insulator, $E^{(in)}$ and $E^{(out)}$ respectively, also can be quite large. The highest

Table 3-2. Electric Fields of the Switch Gap in an Insulating Cylinder.

	Dimensions			Electric Fields			
Case No.	d (cm)	H (cm)	$R/2$ (cm)	$E^{(g)}$ (kV/cm)	$E^{(g)}_{av}$ (kV/cm)	$E^{(in)}$ (kV/cm)	$E^{(out)}$ (kV/cm)
1.	0.5	0.0	4.0	2090	2000	403	422
2.	0.5	2.0	4.0	2090	2000	225	476
3.	2.0	2.0	4.0	770	500	180	375
4.	2.0	2.0	2.0	800	500	820	207
5.	10.0	2.0	4.0	330	100	90	151

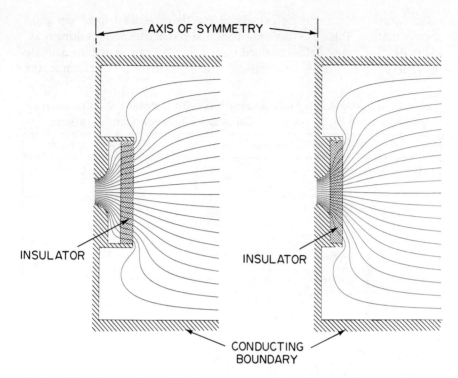

Fig. 3-3. Plots of potentials resulting from voltage applied to electrodes of the switch shown in Fig. 3-2.

fields are associated with the geometry of Case 4, as is evident also from its field distribution seen in Fig. 3-3.

The generic set of switching gaps is used here to illustrate also additional points that can be useful in design of practical switches. Field distributions in Fig. 3-3 show that maximum fields on the outside surfaces of the insulator (components parallel to insulator surface) appear at points where the outer rim of the top (bottom) plates abuts to the insulating cylinder (as in Case 3). These fields can be reduced very substantially by extending the metal rim by a small amount beyond the outer diameter of the insulator. Similarly, the plot of field distribution for Case 4 shifts the parallel component of fields at the surfaces away from the rim-insulator triple point. The ratio of maximum and average fields in the gap ($E^{(g)}/E^{(g)}_{av}$) for Case 5 is much larger than for Case 2, where the aspect ratio H/d is sufficiently small to approximate parallel plate geometry.

The maximum electric fields for some electrode geometries, where end-effects are neglected, can be calculated exactly or at least to a very good approximation. Table 3–3 lists some common geometries and resulting maximum fields. Electric fields obtained from Table 3–3 can be used to estimate the ratio E_{max}/E_{av} for various realistic switch geometries. For example, the

Table 3–3. Maximum Field Strength E with a Potential Difference U between the Electrodes, for Different Electrode Configurations. [*]

Configuration		Formula for F
Two parallel plane plates		$\dfrac{U}{a}$
Two concentric spheres		$\dfrac{U}{a} \dfrac{r + a}{r}$
Sphere and plane plate		$0.9 \dfrac{U}{a} \dfrac{r + a}{r}$
Two spheres at a distance a from each other		$0.9 \dfrac{U}{a} \dfrac{r + a/2}{r}$
Two coaxial cylinders		$\dfrac{U}{2.3\, r\, \lg \dfrac{r + a}{r}}$
Cylinder parallel to plane plate		$0.9 \dfrac{U}{2.3r\, \lg \dfrac{r + a}{r}}$
Two parallel cylinders		$0.9 \dfrac{U/2}{2.3r\, \lg \dfrac{r + a/2}{r}}$
Two perpendicular cylindars		$0.9 \dfrac{U/2}{2.3r\, \lg \dfrac{r + a/2}{r}}$
Hemisphere on one of two parallel plane plates		$\dfrac{3U}{a} \; ; (a \gg r)$
Semicylinder on one of two parallel plane plates		$\dfrac{2U}{a} \; ; (a \gg r)$
Two dielectrics between plane plates ($a_1 > a_2$)		$\dfrac{U\epsilon_1}{a_1\epsilon_2 + a_2\epsilon_1}$

[*]Reproduced from Ref. 3 (following A. B. Bowers and P. Q. Cath, ''The Maximum Electric Field Strength for Several Simple Electrode Configurations,'') *Phillips Tech. Rev.* 6, 270 (1941).

maximum electric field in the gap of the switch shown in Fig. 3-2 (with dimensions listed under Case 5 in Table 3-1) can be approximately compared with the fourth formula in Table 3-3. For potential U of 1 MV, the maximum field in the gap is 540 kV, rather than the more accurate 330 kV obtained from exact calculation.

Other electrode shapes are employed to provide very uniform fields on the electrode surfaces. Reference 20 describes a profile which can be described by a single analytic expression. These expressions can be used for the design of a finite width uniform-field electrode for any desired aspect ratio and field uniformity. Such electrodes are required, for example, for laser cavities and for electron beam controlled volume discharge switches; the latter are discussed in Chapter 9.

An example of a commercially produced switch which demonstrates the use of the design criteria discussed in the preceeding section is a Maxwell Laboratory, Inc. (MLI), high energy spark gap switch, Model 40169. This switch is designed for use in high voltage, high current Marx generators and other applications requiring up to 10 coulombs of charge transfer where no triggering is necessary[21]. Specifications of the switch performance provided by the switch developer, MLI, are given in Table 3-4, the switch is shown in Fig. 3-4.

In designing high power system switching, there arises on occasion a need to combine several gaps in series. One reason for the need may be the practical circumstance that particular components which are available are insufficient for the required voltage hold-off. Another reason for use of series gaps is redistribution of the shock energy that must be absorbed in a given switch. Thus, Neff et al. have developed a 3.5 MV gas switch consisting of 12 gaps capable of conducting currents in the megampere range[22]. Inte-

Table 3-4.   Specifications for MLI Switch, Model 40169.

Maximum operating voltage	120 kV
Minimum operating voltage	60 kV
Maximum peak current	450 kA
Maximum charge transfer	10 C
Inductance	<150 nH
Repetition rate	1 pulse/minute
Environment	Transformer oil
Lifetime	2,000 shots at maximum duty
	10,000 shots at 1 C, 250 kA
Size	10.5″ dia. × 10″ high*
Weight	30 lb*

*English system of measures is retained in this table as given in Ref. 21.

Fig. 3-4. Commercially available spark gap switch, similar in design to an idealized switch shown in Fig. 3-2. Reproduced from Ref. 21, courtesy, Maxwell Laboratories, Inc., San Diego, CA.

grating the gaps, the triggering mechanism, and structural features of the unit, gives an extremely low jitter ($\delta = 1.5$ ns) triggered switch.[22]

The single-gap switch, shown in Fig. 3-4, has no trigger. The multi-gap switch of Ref. 22 has additional electrodes for triggering so that it can initiate discharges within a very short time (< 10 ns) relative to a reference trigger time. Use of additional electrodes that provide triggering can change the field distribution to a degree determined by the position of the trigger electrode relative to the main electrodes. It is possible to arrange such electrodes so that very little field distortion is generated. This is done by use of the mid-plane floating electrode with insignificant effect on the field until a potential is applied to the trigger electrode. Application of the potential causes the field to increase between the trigger and one of the electrodes and to decrease across the remaining gap. The distribution of potential lines associated with the triggered and untriggered cases is shown in Fig. 3-5. This configuration provides one of the very effective methods of triggering. Such a switch configuration is called a "field distortion triggered switch" and its operation is described in the next section.

TRIGGERING

To determine the time of switch breakdown, various triggering methods have been developed for use with spark gap switches. These methods range from relatively simple approaches such as already mentioned; use of the third electrode initially floating (i.e., its potential established by capacitive coupling to the main electrodes) and subsequently connected by some means to ground potential; to more complex approaches involving the use of short-

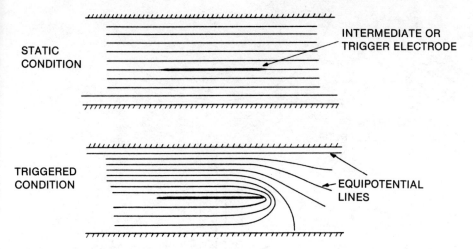

STATIC
CONDITION

INTERMEDIATE OR
TRIGGER ELECTRODE

TRIGGERED
CONDITION

EQUIPOTENTIAL
LINES

Fig. 3–5. Principle of the field distortion switch. Reproduced from Ref. 17, courtesy of U.K. Atomic Energy Authority, Culham Laboratory.

duration lasers which by irradiating the electrodes lead to gap breakdown. In the next chapter, use of ultraviolet radiation for triggering, especially in establishing multi-channel (parallel sparks) discharges, is discussed in relation to low pressure switch operation. In general, gaps become easier to trigger and operate faster the higher the voltage at which they work, and the design task is to prevent them from working in unwanted ways (Ref. 5, p. 4–23). Usually, the choice of the triggering method is determined by the required triggering accuracy (reducing the jitter of triggering delay time) or its complementing property, the range of switching voltage relative to breakdown voltage.

Conventional Triggering

The simplest method of triggering a spark gap switch when accurate timing is not required is to reduce the pressure in the switch once the potential has been applied across the electrodes separated by distance d. Perhaps the most widely used method of triggering is using a trigatron ignition (Ref. 4, p. 221). Figure 3–6 shows schematically a trigatron configuration. One of the two hemispherical electrodes contains a third electrode, the trigger, separated by distance d_{tr} from the main electrode. The purpose of the third electrode is to disturb the electric field applied across the main electrodes to such a degree that breakdown will be initiated. The distortion of the field is achieved by sparking the gap d_{tr} and forming a small discharge. This discharge is accompanied by emission of ultraviolet

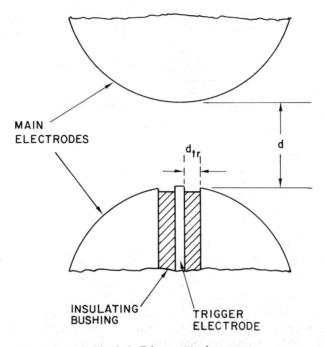

Fig. 3-6. Trigatron spark gap.

radiation (as well as some distortion of the field in the main gap) and a shockwave. As summarized by Komelkov (Ref. 4, p. 183), both of these mechanisms are responsible for the resulting breakdown. T. E. Broadbent's[23] and numerous other works provide evidence that the short-wavelength radiation produces electrons by photoionization which avalanche in the electric field and transform into breakdown channels. A. M. Sletten and T. J. Lewis[24] and R. F. Saxe[25] show that shock ionization of the gas could lead to a breakdown. The accompanying intensification of the electric field in a small region around the trigger electrode serves to provide the necessary condition for avalanching and the resulting breakdown[26].

The trigatron switch is very reliable because of its ruggedness and simple construction. Its erosion rates at high currents are low[11]. Fig. 3-7 shows a fast triggered spark gap[11] assembly operating at 30 kV level at a gap spacing of 10 mm. Such switches have been used in parallel in a 2 MJ bank with a load current capability of 20 MA (at 50 kA, 5 kJ per switch) in a ringing discharge with up to 85% reversal. By choosing an assymetric geometry, as suggested in the patent by L. J. Melhart[8] where the plasma discharge is driven by a magnetic field away from the insulator and from the trigger

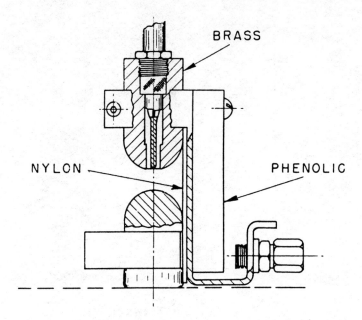

Fig. 3-7. Assembly drawing of the triggered spark gap switch designed for magnetic blow-out. Reproduced from Ref. 11.

pin, a 4000 shot lifetime has been demonstrated. The parallel operation requires that all switching must occur within 45 ns. Figure 3-8, reproduced from Ref. 11, shows the switching time as a function of applied voltage for gap spacing ranging from 8 to 11 mm. At an 8 mm gap setting, the switching time rapidly decreases to about 40 ns as the applied voltage increases to 15 kV and remains nearly constant up to about 23 kV. As the gap spacing is increased, the rapid drop is shifted to higher voltages.

Another triggering scheme frequently utilized is the three-electrode cascade configuration. The control of the cascade or field distortion configuration is achieved through a brief increase in the electric field, between one of the main electrodes and the third, or trigger, electrode to a value considerably exceeding the static discharge field. This can be obtained by applying a trigger pulse or by grounding the trigger electrode through a switch inductance. Figure 3-5 shows schematically the switch with the third electrode being the trigger electrode. When a high voltage trigger pulse is applied to the center electrode, transiently the voltage across one gap decreases and increases across the other gap. The breakdown of the overstressed gap forces the full switch potential to appear across one-half of the original gap resulting in the breakdown of the second half, completing in this manner the closing of the switch.

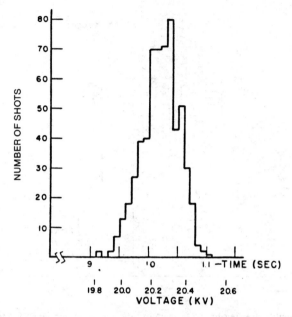

Fig. 3-8. Distribution of time of sparking for 6 mm gap shown in Fig. 3-7. At the low end of the distribution (0.92 s), the mean hold-off time is about 5 s. At the upper end (1.10 s), it is about 0.01 s. The gap voltages shown below the time scale are calculated from the time constant (0.36 s) and power supply voltage (21.5 kV). Reproduced from Ref. 11.

In circuits using the grounding of the trigger electrode, triggering can be achieved rapidly with gap voltage as low as 60% of untriggered gap breakdown voltage (i.e., the switch has a trigger range of 0.6). When the trigger electrode is displaced from the center position, triggering at 40% of breakdown can be achieved (Note 4 in Ref. 5). At tens of kilovolts and above, such gaps stressed at 80% of breakdown level can trigger in a few nanoseconds and have a jitter of few tenths of a nanosecond (Note 4 in Ref. 5). Below 15-20 kV, the triggering operation becomes more difficult due to such factors as the ability to set small gaps sufficiently accurately. A cheap simple gap exists[27, 28] which can be triggered by a 300 volt pulse from high impedance (70 Ω) cable; the trigger pulse is increased to about 1500 V by use of a ferrite-cored transformer.

Numerous practical hints on construction of the gaps and on their operation, including suggestions on how to obtain reliable triggering down to a few kV, has been provided in various discussions by J. C. Martin in Ref. 5, Note 4, and in other references.

A high voltage (60-120 kV) switch for reliable operation has been developed on the basis of the triggering principles illustrated by the configuration in the Fig. 3–9. Maxwell Laboratories' switch model[29] (for triggered and untriggered operation, No. 40113 and 40114 respectively) is shown in Fig. 3–10. The use of air or SF_6 allows up to 1.5 coulombs of charge to be transferred by such a switch in two overlapping regimes. Dry synthetic air (dew point $-60°C$) is used for the 60 to 100 kV range. The use of SF_6 extends the range to 120 kV. Large brass electrodes (including a mid-plane electrode in Model 40113) mounted in cast polyurethane housing provide for long life operation. Static breakdown curves for both types of gas are shown in Fig. 3–11. These curves provide a good illustration of pressure dependence. Table 3–5 lists the switch specifications.

Substantially higher currents, near 500 kA, have been switched by a spark gap switch developed at Culham Laboratories in Great Britain[17]. This switch's main electrodes consist of two replaceable heavy alloy toroidal rings on aluminum switch plates connected directly to a polyethylene insulated transmission line. The 0.6 cm gap pressurized to several atmospheres of air withstands 60 kV. The switch geometry is responsible for the low inductance (15 nH). The MARK III switch shown in Fig. 3–12 has operated for some 2000 shots at 7 C per discharge, without degradation in performance, including the low jitter of ±3 ns (with 22 ns delay), in exhaustive tests. The short delay and low jitter were obtained using the field distortion trigger electrode seen located centrally in Fig. 3–12. This electrode is also made of

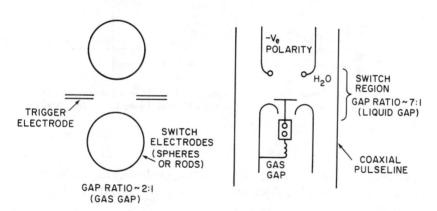

Fig. 3–9. Schematic of field distortion switches using spherical electrodes and planar trigger electrode, typically 1 mm thick. The switch with the planar trigger is shown as a component of a coaxial transmission line. For a hole in the positive electrode with a diameter of the order of gap separation, used for introducing the trigger electrodes, small errors in the trigger voltage do not have any effect on self-breakdown voltage (J.C. Martin, Note 4, Ref. 5). When the trigger electrode is displaced so that gaps are in the ratio of 1:2, both gaps break down simultaneously.

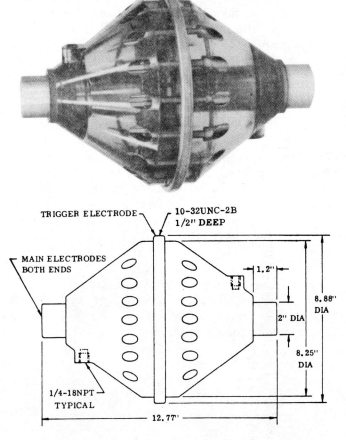

Fig. 3-10. High energy spark gap switch manufactured by Maxwell Laboratories, Inc., and its schematic. Reproduced from Ref. 29, courtesy Maxwell Laboratories, Inc., San Diego, CA.

heavy alloy. It is smaller in diameter than the circle defining the minimum main electrode separation. Such an arrangement prevents the arc from contacting the main switch insulation.

The triggering mode of the switch in Fig. 3-12 is a two-stage cascade breakdown with negative voltage of 30–60 kV applied to the trigger electrode. This causes the first stage of breakdown to occur between the trigger and the high voltage positive electrode. The second stage of breakdown occurs as the trigger electrode approaches the potential of the high voltage electrode. The first stage gap must be larger than the second gap to produce good triggering.

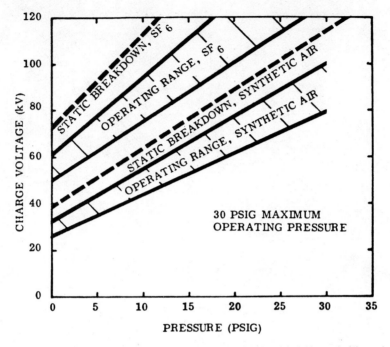

Fig. 3–11. Static breakdown curves and operating range for the switch shown in Figure 3-10. Reproduced from Ref. 29, courtesy, Maxwell Laboratories, Inc., San Diego, CA.

The precise triggering achieved in the MARK III switch makes it practical to use in connecting parallel capacitors to a common load without use of any significant isolation leads. The high current and large charge transfer capability of the switch (e.g., in comparison with the capability listed in

Table 3–5.   Mark III Switch.

Maximum operating voltage	120 kV
Maximum peak current	150 kA
Maximum charge transfer	1.5 C
Inductance	<100 nH
Jitter	<50 ns
Switching Delay time	50 ns
Trigger amplitude	120 kV
Trigger rate of rise (minimum)*	5 kV/ns
Repetition rate	1 pulse/min.
Environment	Transformer oil
Dielectric gas	Synthetic air or SF_6
Lifetime	10,000 shots at maximum duty, Longer at reduced duty

*For lowest jitter

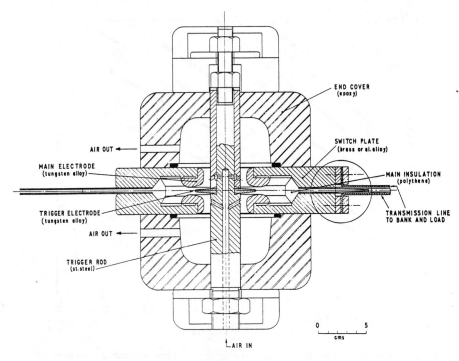

Fig. 3–12. MARK III high current triggered spark gap switch. Reproduced from Ref. 17 (Report CLM-R71), courtesy of U.K. Atomic Energy Authority, Culham Laboratory.

Table 3–5) makes it a good candidate for those applications where the number of switches must be kept small.

Laser Triggering

The invention of the laser has provided the switch designer with yet another method for precision triggering of high power switches. Very high power density associated with focused laser beams led to early attempts in 1965 by a group at the Air Force Weapons Laboratory[30] and by others[31, 32, 33] to exploit lasers for switch triggering. This was done keeping in mind two important advantages. First was the possibility of very precise (really simultaneous) triggering of many parallel gaps to obtain an equal sharing of current in large pulsed energy storage systems, such as the PBFA I system at Sandia Laboratories[34]. The second anticipated advantage was the elimination of the need to use trigger cables in high voltage systems where use of such cables is very impractical, for example, due to severe maintenance requirements.

Typically, a laser-triggered switch consists of a gap with an aperture of a few millimeters in diameter in the anode for passage of a laser beam. The beam is focused in the gap by a lens having an appropriate focal length to provide the highest power density of the laser beam on the cathode surface. (As the focal spot is moved away from the cathode, the laser power required to initiate breakdown, with a given delay relative to the laser pulse, increases rather strongly[33]). Surface plasma formed at the cathode expands and distorts the electric field applied across the gap leading to a breakdown, and provides a source of ultraviolet radiation (and effectively ionizes the gas within 1 cm of the cathode). The laser beam power (using ruby lasers) required for the triggering of the switch by this mode is in the range of 1–10 MW.

Another mode that can be used for triggering a switch by a laser beam is to introduce the beam transversely to the axis of the gap, focusing it at a point in the gas somewhere between the electrodes. A sufficiently high power density in the focus will cause the gas to break down, forming a plasma of a few millimeters diameter. The plasma's high conductivity leads to breakdown of the gap by distorting the applied field across the gap in a manner similar to the surface-irradiated switching. Again, the ultraviolet emission from the gas breakdown plasma also contributes to gap breakdown. The mechanisms of gas breakdown and subsequent discharges are described fully in Chapter 7 ("Irradiation and Time Lags") of Ref. 12. Chapter 9 ("Laser Induced Electrical Breakdown of Gases") of the same reference discusses further the laser interactions with the gas leading to breakdown in localized regions. Breakdowns forming very long channels, of the order of 1 meter in length, have been studied by R. J. Greig et al.[35] Again, as in the case of electrode irradiation, few MW laser beams of short duration focused to produce about 10^{10} to 10^{11} W/cm^2 are required to produce gas breakdown.

Precise control of switch gap breakdown by laser triggering led to the investigation and detailed characterization of several switch designs for pulse power applications. R. G. Adams et al.[34], for example, investigated a 2.8 MV gas switch insulated with SF_6. Because of the large separation between the electrodes, they found normally used visible or IR laser irradiation of the switch electrode[30] to be insufficient for the operation of the high voltage gap. Using an ultraviolet beam of less than 0.1 J from a KrF oscillator-amplifier laser system, they obtained 0.5 μs jitter in triggering for switch voltages of 70–90% of the self-breakdown voltage. Such a broad operating range resulted from use of the laser beam to ionize the entire 11 cm channel between the electrodes. To obtain very low inductance switching, they also used ultraviolet laser beams to obtain multichannel breakdown in the rail gap switches shown in Fig. 3-13[36]. A few-mJ KrF laser beam of 10 ns du-

(a)

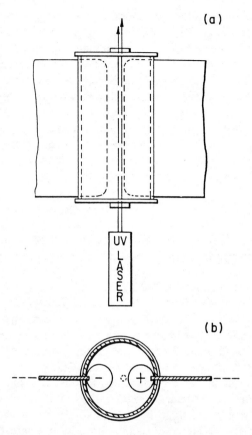

(b)

Fig. 3-13. Two views of the laser-triggered low inductance multichannel rail gap switch. Reproduced with permission from Ref. 36. © 1982 IEEE.

ration is shown to be directed along the surface of the long (positive) electrode. Use of pure Ar gas in the switch resulted in very good multichanneling. Such multichanneling was also obtained with much higher voltage hold-off (of 40 kV/cm at 1 atm) when a gas mixture of 1% SF_6, 50% N_2 and 49% Ar was used. Fifty to seventy channels per meter of the electrode length produced sub-nanosecond jitter in triggering time (with 0.25 ns delay time[36]).

Finally, L. L. Hatfield et al.[37] demonstrated that it is also practical to use fiber optics for triggering gas gaps with low jitter. This approach permits more flexibility in integrating precisely triggered switches into complex pulse power generators.

ELECTRODE EROSION

The erosion of the spark gap electrodes results from the heating of their surfaces by the heat flow from the discharge channel. Various heat transport mechanisms take part in this process, among them Joule heating near the electrode surface by the current feeding the discharge channel, the thermal motion of the discharge plasma particles, absorption of light generated in the channel, and absorption of the kinetic energy of accelerated ions and electrons striking the negative and positive electrodes respectively. The last process is often the most dominant. The amount of heat in high power switches is sufficient to melt and/or vaporize the surfaces of the electrodes at a rate which also depends on the electrodes' thermal conductivity. Examination of the electrodes after discharging shows an unambiguous erosion of the surface from melting and subsequent motion (due to large pressure at the roots of the discharge channel), as well as removal of the material from the electrodes due to vaporization or ejection of solid and liquid material. In summary, the rate of erosion depends on the thermal conductivity of the electrodes, on the electrode material, and on the behavior of the discharge channel. The effects of erosion are important in determining the life and performance of gap switches. Various electrode phenomena associated with the presence of arcs is discussed and well referenced by R. Holmes in Chapter 11 of Ref. 12. The choice of best materials for the electrodes has also been studied extensively; much of that work is referenced by A. L. Donaldson et al. in two companion articles[38, 39].

A. L. Donaldson et al.[38, 39] have measured erosion rates, i.e., the rate of loss of material from the electrodes for various materials at levels of current that are of interest to high power switching (i.e., in the range of 10–800 kA). Not unexpectedly, they have confirmed that the electrode material with the lowest erosion rate is copper-tungsten 3W[40] when operated in a N_2 atmosphere for atmospheric pressure switches. Other copper-tungsten alloys are nearly as good. Stainless steel alloys have two to three times higher erosion rates and such specialized materials as copper-graphite (e.g., type DFP-1C[41]) or pure graphite have about ten times greater erosion rate than the best tungsten alloy. The rate of erosion, which is linear, depends primarily on the amount of charge transferred between the electrodes. Donaldson et al.[38, 39] gives the rates for copper-tungsten alloy of about 12 to 20×10^{-9} m^3 as the charge transfer increases from 10^4 to 2×10^4 C. In their experiments, the charge was accumulated over some 50,000 discharges. The dependence of the erosion rate on the charge transfer is the most important switch design factor affecting the reliability and life of the switch. (However, some marginal effects on the erosion rate due to much weaker dependence of erosion on current amplitude, gases in the gap region, and on pulse duration can also be expected.)

ILLUSTRATION OF SWITCH INTEGRATION INTO PULSER SYSTEM

The diverse set of considerations required to design gas gap switches for high power applications have been collected in this chapter. This diversity is typical of the information required in the design of switches discussed in the succeeding chapters. An additional set of concerns arises when such switches are to be incorporated into large pulser systems, with the switch characterization as a circuit component discussed in Chapter 2 being one of the key concerns. Because gas gap switches are the most widely used switches, an example of their use is provided. We consider the high current (150 kA) TRITON accelerator constructed at the Naval Research Laboratory to produce beams of relativistic electrons with energies of 1.2 MeV[42]. The incorporation of the gas switches as integral parts of the TRITON pulser shows how pulser considerations and requirements influence the evolution of the specialized design of the switches. This example can serve, further, to guide the reader in the design of new high power systems or in upgrading the existing systems. While the configuration discussed here concentrates on the large switching elements (operating at 1 MV level) which are integrated into the Blumlein pulse-forming section of the accelerator, with their location shown in Fig. 3–14, it should be noted that some twelve smaller gas gap switches (operating at up to 85 kV) are utilized in the primary storage Marx capacitor bank of the accelerator[42].

To carry out the scientific program served by the TRITON accelerator, the pulse generator must be command-triggered to correlate the electron beam injection with formation of the external theta pinch plasma and with a Q-switched laser used for light scattering diagnostics. These requirements have introduced special problems in the design of the pulse generator as well as the switching. The accelerator shown in Fig. 3–14 consists of an 8 ohm transmission-line pulse generator using water as a dielectric. It is pulse-charged in about 1 μs by a Marx capacitor bank and discharged into an electron beam diode. The 12 switches connecting the charged 0.5 μF capacitors of the bank in series are housed in a single long switch column. The gaps are 1.3 cm long and can be pressurized to 2 atm of N_2. The first switch is triggered and the remaining ones close due to resulting overvoltage. Switch jitter within acceptable limits by the irradiation of the opened gaps by those already closed. Variation of pressure from 0.2 to 2 atm allows the Marx generator to operate with capacitor charging voltage of 40 to 85 kV, corresponding to an output voltage of 0.5–1.0 MV. To form the short duration pulse, the pulse forming section uses an 8 ohm transmission line in the Blumlein configuration because it requires only one-half the voltage compared to that which is needed by a simple transmission line to produce the same output voltage. The Blumlein charged by the Marx is discharged to form a high power pulse of 75 ns in duration by four trigatron switches

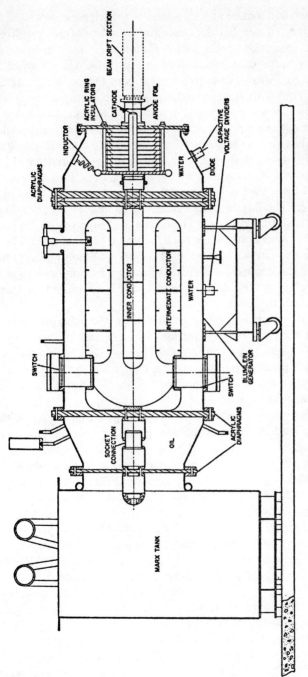

Fig. 3–14. Cross-section view of the TRITON accelerator using a Blumlein transmission line as a pulse-forming generator. Reproduced from Ref. 42. © 1985 IEEE.

placed in the outer coaxial line for access of the trigger cables. The location of the switches require that the generator must be charged positively.

The pulse generator uses a folded Blumlein circuit consisting of three coaxial cylindrical conductors as seen in Fig. 3–14. The intermediate conductor is charged with respect to the other two. Burton et al.[42] describe the two coaxial lines with respect to optimizing the energy storage. The lines formed in this manner are charged in parallel by the Marx bank. The output pulse is formed by short-circuit switching of the outer transmission line, connecting the two lines in series for discharge into the load at voltage level consistent with twice that of the single line voltage referenced with respect to the voltage across an open circuit load. The pulse generator outer coaxial line is 1.8 m long, 0.65 m in diameter, and contains 2500 liters of demineralized water. The intermediate cylindrical conductor is 1.25 m long with a one way transit time of 37 ns. The mechanical support for this conductor is provided by the switches at the back (nearer to Marx tank) and by a single nylon rod hanger (shown to the left of the acrylic diaphragms) in the front and the electrical connection is made through a Lucite diaphragm (at left) and an oil-filled section. The center cylinder is supported by Lucite diaphragms at each end of the cylinder.

The pressurized gas trigatron was chosen for the pulse generator switching to allow for command triggering with sufficient precision so that all four parallel switches could close simultaneously (i.e., within the switch-to-switch transit time of about 10 ns). This choice was made in lieu of a commercially available 1 MV switch using SF_6 gas[43], because the test showed that the trigatron using such a gas will operate properly only if the electrode with the trigger pin is positive. Because of the positive polarity on the other electrode required by the intended application of the generator, the switch would not perform adequately. Although the N_2-filled switch would operate properly, the lower hold-off voltage required that a small amount of SF_6 be mixed with N_2. Addition of up to 10% of SF_6 to a 2 cm test gap showed that shot-to-shot jitter remained acceptable, and scaled to larger gaps as well for trigger signal of 10 kV/ns.

Fig. 3–15[42] shows a cross-section view of the 1 MV switch and its assembly into the outer part of the Blumlein generator. This switch design resembles the generic switch type shown in Fig. 3–2. The housing is MC-901 nylon tube with a 2.5 cm thick wall. The end-plates are made of stainless steel and attached to the tube using some 30 screws and "Heli-coil" inserts in the nylon. The design is sufficient to withstand a pressure of 15 atm. The electrodes (separated by 5 cm) are made of brass; the trigger signal is brought on a high voltage RG-220/U cable whose center conductor is inserted through the outer electrode with the solid copper conductor performing as the trigger pin. The outer braid of the cable is cut back several centimeters

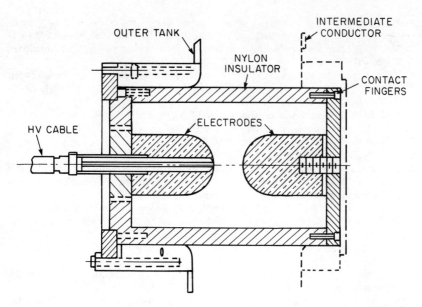

Fig. 3–15. Cross section view of the high voltage switch as it appears when assembled into the outer transmission line shown in Fig. 3–14. Reproduced from Ref. 42 by permission of J. K. Burton (first author) of Naval Research Laboratory. © 1985 IEEE.

and clamped to the electrode to provide the trigger signal return path. The total switch inductance is about 170 nH. Installation of this switch in the TRITON accelerator has resulted in switching performance with a simultaneity of a few ns. The current has been observed to be shared equally among the four switches.

This example of the development of a switch in close relationship to pulser design demonstrates the degree of interaction between various design constraints. Successful solution of the problems posed by these constraints has resulted in production of intense electron beams with power of 10^{11} W for studies of plasma heating processes in various scientific programs.

REFERENCES

1. *IEEE Conference Record of 1982 Fifteenth Power Modulator Symposium,* IEEE Catalog No. 82 CH1785-5, (1982). Earlier Modulator Symposium Proceedings also contain relevant material.
2. *Third IEEE International Pulsed Power Conference,* IEEE Catalog No. 81 CH1662-6 (1981). Earlier Pulsed Power Conferences proceedings also contain relevant material.
3. T. R. Burkes, "A Critical Analysis and Assessment of High Power Switches," Naval Surface Weapons Center (Dahlgren, VA), Report NP 30/78 (1978).

4. V. S. Komelkov, "Technology of Large Impulse Currents and Magnetic Fields," translated by Foreign Technology Division, Wright-Patterson AF Base, Ohio, Report FTD-MT-24-992-71, from *Tekhnika Bolshikh Impulsnykh Tokov i Magnitnykh Poley,* Atomizdat, Moscow (1970).

5. C. Baum, ed., "Pulsed Electrical Power Circuit and Electromagnetic System Design Notes, PEP 4-1," Air Force Weapons Laboratory (Albuquerque, NM), Report AFWL-TR-73-166 (1973).

6. A. J. Buffa and Sol Schneider, U.S. Patent 3,353,059 (Apr. 8, 1966).

7. J. L. Harrison, U.S. Patent 3,320,478 (May 16, 1967).

8. L. J. Melhart, U.S. Patent 3,267,320 (Jul. 30, 1962).

9. G. C. McFarland, U.S. Patent 3,312,368 (Apr. 23, 1963).

10. J. Harrison, R. White, V. G. Chen, T. Day, D. Husovsky, and S. Lloyd, "A Transportable Simulation Pulser," *Proceedings of the Fifth Symposium and Technical Exhibition on Electromagnetic Compatibility,* ETH Zentrum–IKT Zurich (1983).

11. W. H. Lupton, "Fast Triggered Spark Switches for a Two-Megajoule Capacitor Bank," *Proceedings of Fifth International Conference on Ionization Phenomena in Gases, Munich, 1961,* North Holland, Amsterdam, p. 2059 (1961).

12. J. M. Meek and J. D. Craigs, eds., *Electrical Breakdown of Gases,* John Wiley & Sons, New York, p. 210 (1978).

13. F. L. Jones, "Ionization Growth and Breakdown," *Encyclopedia of Physics,* Vol. 22, Springer-Verlag, Berlin, pp. 1–52 (1956).

14. J. C. Martin, "Nanosecond Pulse Techniques," AWRE (Aldermaston, U.K.) SSWA/JCM/704/49 (1970).

15. P. Felsenthal and J. M. Proud, "Nanosecond Pulse Breakdown in Gases," Technical Report No. RADC-TR-65-142, Rome Air Development Center, Griffins AFB, NY, (1965).

16. L. S. Levine and I. M. Vitkovitsky, "Pulse Power Technology for Controlled Thermonuclear Fusion," *IEEE Trans. on Nucl. Sci.* NS-18, 255 (Aug. 1971).

17. A. E. Bishop and G. D. Edmonds, "A 100 kV Low Inductance Switch (Spark Gap) for Starting Diverting and Charging Capacitor Discharge," Culham Laboratory (Abingdon, Berkshire, U.K.) Report CLM-P81:21 (1966) and P. M. Barnes, J. E. Gruber, and T. E. James, "High Current Switching Using Low Inductance Field Distortion Spark Gaps Closely Connected in Parallel," Culham Laboratory (Abingdon, Berkshire, U.K.) Report CLM-R71 (1967).

18. T. H. Lee, *Physics and Engineering of High Power Switching Devices,* MIT Press, Cambridge, MA. 1975.

19. J. D. Shipman, unpublished computer program developed at Naval Research Laboratory, Washington, D.C. The program is based on J. E. Boers, "FEEARS—A Digital Computer Program for the Simulation of Laplace's Equation Including Dielectric Interfaces and Small Ungrounded Electrodes," Sandia Laboratory Report SC-RR-71 0377, Albuquerque, NM (1971).

20. T. Y. Chang, "Improved Uniform-Field Electrode Profiles for TEA Laser and High Voltage Applications," *Rev. Sci. Inst* 44, 405–407 (1973).

21. Maxwell Laboratories Engineering Bulletin, EB 1200, MLI, 8835 Balboa Ave., San Diego, CA 92123 (1981).

22. T. Neff and G. Frazier, private communication, Physics International Co., San Leandro, CA, 1982.

23. T. E. Broadbent, "The Breakdown Mechanism of Certain Triggered Gaps," *Brit. J. Appl. Phys.* 8, 37 (1957).

24. A. M. Sletten and T. J. Lewis, "Characteristics of the Trigatron Spark Gap," *Proc. Inst. Electr. Eng.,* Pt. C., 104, 54 (1957).

25. R. F. Saxe, "The Breakdown of a Triggered Spark Gap," *Proceedings of Fifth International Conference on Ionization Phenomena in Gases,* Vol. 1, North-Holland Publishing Co., Amsterdam (1962).

26. G. Goldenbaum and E. Hintz, "Pressurized Trigatrons with a 10 kV-50 kV Low Jitter Operating Range," *Proceedings of the Sixth International Conference on Ionization Phenomena in Gases,* Vol. 2, p. 331, Paris (1969).

27. J. C. Martin, U.K. Patent 1,080,211 (1970).

28. R. J. Rout, "Triggered Spark Gaps for Image Tube Pulsing," *J. Sci Instrum. (J. of Phys. E)* Ser. 2, Vol. 2 (8), 738–739, (1969).

29. Maxwell Laboratories Engineering Bulletin EB 1025, MLI, 8835 Balboa Ave., San Diego, CA (1982).

30. W. K. Pendelton and A. H. Guenther, "Investigation of a Laser Triggered Spark Gap," *Rev. Sci. Inst.* **36,** 1546–1550 (1965).

31. J. J. Bleeker and C. G. Morgan, "Some Design Considerations of a Laser-Triggered Impulse Generator," CERN Report NAP/INT 65-31 (1965).

32. F. Deutsch, "Triggering of a Pressurized Spark Gap by a Laser Beam," *Brit. J. Appl. Phys. (J. Phys. D),* Ser. 2, Vol. 1, 1711–1719 (1968).

33. A. I. Babalin, B. I. Petrov, V. A. Rodichkin, and A. M. Timonin, "Laser-Triggered Spark-Gap Switch," *Sov. Phys. Tech. Phys.* **15,** 1335–1338 (1971).

34. R. G. Adams, J. R. Woodworth, and C. A. Frost, "UV-Laser Triggering of Multi-Megavolt Switches," *IEEE Conference Record of the Fifteenth Power Modulator Symposium,* IEEE Cat. No. CH1785-5, pp. 79–83 (1982).

35. R. J. Greig, P. W. Koopman, R. F. Fernsler, R. E. Pechacek, I. M. Vitkovitsky, and A. W. Ali, "Electrical Discharges Guided by Pulsed CO_2-Laser Radiation," *Phys. Rev. Lett.* **41,** 174 (1978).

36. R. S. Taylor and K. E. Leopold, "UV Radiation Triggered Rail Gap Switches," *IEEE Conference Record of the Fifteenth Power Modulator Symposium,* IEEE Cat. No. 82CH1785-5, pp. 113–117 (1982).

37. L. L. Hatfield, H. C. Harjes, M. Kristiansen, A. H. Guenther, and K. H. Schonbach, "Low Jitter Laser Triggered Spark Gap Using Fiber Optic," *Digest of Technical Papers of the Second IEEE International Pulsed Power Conference,* Lubbock, TX, IEEE Cat. No. 79CH1505-7, pp. 442–445 (1979).

38. A. L. Donaldson, M. O. Hagler, M. Kristiansen, G. Jackson, and L. Hatfield, "Electrode Erosion in a High Energy Spark Gap," *Digest of Technical Papers of the Fourth Pulsed Power Conference,* IEEE Cat. No. 83CH1908-3, pp. 530–533 (1983).

39. A. L. Donaldson, M. O. Hagler, M. Kristiansen, G. Jackson, and L. Hatfield, "Electrode Erosion Phenomena in a High Energy Pulsed Discharge," *IEEE Trans. on Plasma Science* **PS-12,** 28–37 (1984).

40. Contacts Metals Welding, 70 South Gray St., Indianapolis, IN 46201.

41. Poco Graphite, 1601 South State St., Decatur, TX 76234.

42. J. K. Burton, J. J. Condon, M. D. Javnager, W. H. Lupton, and T. J. O'Connell, "The TRITON Electron Beam Accelerator," *IEEE Proceedings of the Fifth Symposium on Engineering Problems of Fusion Research Conference Record,* IEEE Cat. No. 73CH0843-3NPS, pp. 613–618 (1973).

43. D. Markins, "Command Triggering of Synchronized Megavolt Pulse Generators," *IEEE Trans. Nucl. Sci.* **NS-18,** 196 (1971).

4
Vacuum and Low
Pressure Discharges

INTRODUCTION

The superior insulating properties of vacuum (and gasses at low pressures) have led to the development of switches operating in this regime. One obvious reason for considering vacuum switches is the absence of strong shocks generated by discharge channels at atmospheric and higher pressures. Forces in a vacuum switch are dominated by magnetic forces caused by the switch current and transient overpressure resulting from electrode and insulating wall erosion, both of which can be significant when high energy and power densities are switched. The need for high rate repetitive operation also provides a reason for developing a variety of vacuum and low pressure switches[1, 2, 3]. Vacuum switches have two properties, high recovery rate and low erosion rate, which are needed for such operation[1]. In addition, vacuum switches are desirable when the load is also a vacuum element such as a vacuum inductive storage line or an intense charged particle beam diode. When vacuum switches are used with other vacuum elements, consideration must be given to the need for vacuum pumps and seals to maintain the proper pressure. The turn-on time for vacuum switches is generally longer (about 0.1 μs) compared to that of high pressure gaps; jitter can be in the nanosecond range[1, 3].

Prior to a discussion of these switches, the definition of "vacuum" and "low pressure" as used in this book should be considered. The gas density level is called a "vacuum" when the mean free path of the molecules is greater than that of the switch cavity. The mean free path is, for example, about 1 cm at a pressure of 10^{-2} torr (corresponding to about 3×10^{14} molecules/cm^3). As the mean free path becomes substantially smaller then the cavity dimensions, but less then at atmospheric pressure, the switch is said to operate at "low pressure". This distinction is important in considering triggering and conduction mechanisms. For instance, the use of electro-negative gases in the vacuum regime would not be expected to contribute significantly to recovery. However, when the molecular density is sufficiently high, such gases interact with free electrons, thus affecting the switch operation.

There are two types of vacuum and low pressure switches. The first type is essentially a spark gap, usually triggered, which relies on low pressure for voltage hold-off. This type of switch is relatively widely used with at least two commercial devices introduced on the market[1, 3]. The growing use of such switches is also reported in the USSR[4]. The second type is a surface flashover switch with its properties determined by the dielectric surface and electrode configuration. Specific studies have been made in each case to reduce the inductance, electrode or dielectric surface erosion, and reliability by distributing the current density over large electrode areas. In the first case, the so-called pseudo-spark switch has evolved from the low pressure spark gap switches to distribute the current density away from the high electric field enhancement points[5]. In the case of surface flashover switches, attempts to induce the discharges into multiple channels to reduce the concentration of current density (in order to minimize erosion and damage at electrodes) have been successful. Some of these designs have been employed to switch currents of several megamperes[3, 6].

A derivative of the vacuum closing switch is the so-called metal plasma arc switch (MPAS)[7]. As in other low pressure switches, the electrode material plays an important role in the closing of the switch and subsequent conduction. However, these arcs are controlled either by the external magnetic fields or electrode configuration to provide current interruption as well. This can be done at relatively low power only and thus is not considered further here or in Chapter 7 dealing with current interruption.

The fast recovery of the low pressure switches requires the same characteristics as those which must be satisfied by the vacuum circuit breakers discussed in Chapter 7. Thus, many design considerations are similar and the references given in Chapter 7 should also be consulted in choosing the vacuum or low pressure switches for closing of high power pulse generation. The usual voltage stand-off is in 50–100 kV range. Because the spontaneous breakdown of the gap is much less sensitive to pressure variations for small electrode spacings[1], series connection of many gaps to develop high voltage switches in a megavolt regime have been used[3, 8].

The results of vacuum switch breakdown studies have also been applied in the design of intense electron beam diodes, especially in pulsed operation and also where "magnetic" insulation (as discussed in Chapter 9) plays a role.

BREAKDOWN

Vacuum and Low Pressure Gaps

Untriggered Initiation of Breakdown.   The breakdown of vacuum and low pressure gaps has been studied at least since the 1920's[9]. Applications have ranged from the design of dc x-ray tubes to pulsed intense charged particle beam diodes to high voltage components on spacecraft operating outside the earth's atmosphere. Reviews of this work and of current understanding are given by Chatterton in Chapter 2 of Ref. 10, Latham[11], Farral in Chapter 2 of Ref. 12, Litvinov[13], Noer[14], Hinshelwood[15], and in the references therein.

Breakdown (at least) under pulsed conditions occurs by a mechanism known variously as *field, explosive,* or *cold cathode emission.* The application of an electric field across the gap results initially in a flow of electrons from localized regions on the cathode, with high current density but low total current. At high field strengths, the local energy density due to this current is sufficient to vaporize and ionize the cathode material. The resulting plasma flares expand and merge to form a surface from which large macroscopic currents may be drawn. The cathode plasma layer also expands toward the anode, eventually shorting the gap. The plasma formation time is inversely related to the applied field strength, being typically on the order of a ns for fields on the order of 1 MV/cm. Gap closure occurs at several cm/μs. The exact breakdown mechanism(s) are still a subject of some controversy.

Electrons in a (cold) metal are prevented from escaping into the vacuum by a potential step (the work function); the presence of an electric field changes this step into a barrier of finite width so that quantum mechanical tunneling may take place. Electron emission from cold surfaces is described by the Fowler-Nordheim equation[16]. If the emission current density is sufficient to heat the metal, then thermionic emission, in which a fraction of the electrons has enough thermal energy to surmount the barrier, can also take place. Pulsed breakdown of vacuum gaps, however, occurs at field strength 1–3 orders of magnitude below that required by the basic theories cited above. Until recently, this discrepancy was accounted for by the assumption that breakdown originates at metallic microprotrusions ("whiskers") on the cathode surface, where geometric field enhancement would be invoked to yield the necessary (microscopic) electric field. The high local current density would lead to a regenerative thermal instability, resulting in an explosion of the whisker and subsequent plasma flare formation.

The results of more recent experiments, however, have suggested that dielectric surface contaminants play a significant role in vacuum breakdown, at least in certain regimes. Several theories, involving different types of contaminants (inclusions, submicron concentrations, thin absorbed films, etc.) have been proposed, and they are described in the above-mentioned reviews, with that of Noer[14] being the most comprehensive. None of the (metallic or nonmetallic) mechanisms so far proposed can explain all of the data, and it is likely that more than one mechanism may be involved, especially upon consideration of the wide range of experimental conditions (surface roughness, cleanliness, etc.).

Under dc or long ($>\mu$s) pulsed conditions, additional mechanisms can contribute to breakdown. The factor differentiating the behavior of dc and pulsed breakdown is the time required for various mechanisms to operate relative to the risetime of the applied voltage pulse. Microscopic electron emission from the cathode is still presumably the initiating factor. This emission can vaporize (or outgas) regions of the anode leading to accumulation of anode material into the gap. In addition, microparticles from either electrode may become detached under the action of the current or field and cross the gap (after acquiring a surface charge), striking the opposite electrode and releasing more material, and so on.

Regardless of the microscopic breakdown mechanism(s), there is a wealth of experimental data regarding the effect of various electrode parameters on breakdown strength. (Much of this is discussed in the reviews cited above.) In using this data for switch design, it is important to take into account the relevant experimental conditions, such as vacuum quality or pulse-length. To obtain a more quantitative picture of the breakdown, the initiating process, field emission of electrons, is considered following the Fowler-Nordheim idealized framework, with an observation that the emission of electrons from metals occurs at temperatures below 1000°K and consists almost entirely of the electric field emission[12]. The emission current can be derived by considering the barrier penetration probability and the source function (calculated from the Fermi-Dirac equation and from an estimate of the barrier model)[10]. The current density from cold surfaces is calculated quantum-mechanically to yield the Fowler-Nordheim equation. In the mks system, this equation is[12]

$$J = \frac{1.541 \times 10^{-2}E^2}{\psi t^2(y)} \; \exp\left[\frac{-6.831 \times 10^9 \psi^{3/2} v(y)}{E} \right] \quad (4\text{-}1)$$

with

$$y = \frac{3.795 \times 10^{-3}\sqrt{E}}{\psi} \quad (4\text{-}2)$$

where J is current density in A/m^2, E is the electric field in V/m, and ψ is the work function in eV. Although the functions $t(y)$ and $v(y)$ have been tabulated[17], they can usually be approximated as constants. In that approximation, the plot of $\ln (J/E^2)$ versus $1/E$ is a straight line with a negative slope. A numerical example of the field emission current density for $\psi = 2.5$ and 4.5 eV for a radius of curvature giving a field enhancement of 100 is presented as Fig. 2.6 in Ref. 10.

Following the "whisker" model, which postulates high electric fields at tips of metallic whiskers on electrode surfaces, the observed whisker diameters suggest that it is possible to reach emission current densities of 10^{12} A/m^2. A typical value of the electric field for tungsten whiskers is 6×10^9 V/m (Ref. 12). At this current density, sufficient Joule heating of the tip occurs, resulting in the vaporization of the emitter, injecting higher density gas into the gap and leading to ionization and breakdown[18]. Under different conditions, for example when the electrode separation is large, another mechanism can also operate. It has been postulated[19] (and confirmed by studies of luminosity at the anode during early breakdown stages[20, 21]) that electron emission from high field enhancement (accelerated to higher voltages in longer gaps) strike the anode surface and release a cloud of positive ions. The cloud of ions at the cathode further concentrates the field enhancement, leading to a breakdown.

Breakdown voltage of vacuum gaps for dc conditions is given in Fig. 4–1, adapted from the data of Smith et al.[22] Fields of 200 kV/cm, substantially higher than those in atmospheric air, can be supported across copper electrodes by vacuum (pressure $< 10^{-6}$ torr) with a gap separation of 1 cm. However, there are large differences in the hold-off field when different metals are used. Figure 2.1 in Ref. 10 shows these differences for electrode separations from less then 10^{-2} cm to about 1 cm. At 0.1 cm, the vacuum between lead electrodes holds-off approximately a three times lower field than in the case of steel electrodes. Aluminum and copper hold-off is intermediate between the steel and lead. This difference is determined by a number of surface properties of metals: rate and type of oxide formation, maximum outgassing temperature, history of thermal cycling, impurity content, and other factors. M. Okawa et al.[23a] attempt to systematize the vacuum breakdown for two types of electrodes by developing scaling relations for the breakdown field, E_b, as a function of the electrode surface area, S_{eff}. They define S_{eff} as the area where the electric field is within 10% of its peak value. These scaling relations are analogous to those for gases, liquids, and solids presented in Chapter 3 (Eq. 3–3), Chapter 5 (Eq. 5–1) and in Chapter 6 (Eq. 6–2) respectively. According to Ref. 23a, $E_b = kS_{eff}^{-0.24}$ in kV/cm for S_{eff} in cm^2. The value of a constant k is 380 for copper and 580 for stainless steel electrodes. This scaling was tested in the range 100 kV/cm $< E_b <$ 1000 kV/cm, with electrode areas limited to < 100 cm^2. The data in Ref. 23a indicates that the absolute value of the area exponent decreases for $S_{eff} > 100$ cm^2.

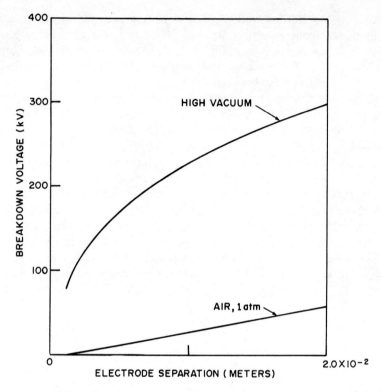

Fig. 4-1. Comparison of the dc breakdown voltage of atmospheric pressure and vacuum gaps for small separation of electrodes.

One important factor is the effect of electrode conditioning, that is, the exposure of electrodes to some type of discharges (the so called "spark" conditioning[11]) or to ion bombardment[12]. In the discussion of vacuum switches later in this chapter, the need of conditioning for specific switches is emphasized. Such conditioning may also be affected by introducing low pressure gases into the switch[10, 11, 12]. At pressures of 10^{-3} torr of argon, it is possible, for instance, to run a low current (tens of microamperes) prebreakdown discharge which forms positive ions generated by the electron impact in the high field region; the positive ions in turn accelerate toward the microprotrusions (causing the emission) to blunt their high field enhancement. In some experiments[23b], breakdown fields can be improved by such techniques by a factor of 3 to 5. In other experiments, such improvement may be obscured by the effect of microparticles on breakdown.

As the pressure between the electrodes is increased, not just for conditioning purposes but for switching operation (e.g., so that simpler pumping can be used), from a vacuum to about 10^{-4} torr, the hold-off improves substantially, as can be seen in Fig. 4-2, reproduced from Ref. 24, as published in Ref. 12. This enhancement ends abruptly between 10^{-1} and 10^{-2} torr for gap separations between 0.1 to 2.5 cm, getting into the pressure region associated with the branch of the Paschen curve, where the collisional effects become important. The low pressure side of the maximum in Fig. 4-2 appears to be determined by the reduction in the number of ions bombarding the negative electrode, since the pressure is reduced[12].

Other more subtle effects then the electrode conditioning also are associated with the introduction of gases into the inter-electrode region, especially as they affect dc operation of such switches. One of these effects, the predischarge or "dark" current, although low, can lead to erratic breakdown or affect pulser systems using vacuum switches in some undesirable way. Thus, when carefully conditioned molybdenum electrodes are used, the dark emission current is not eliminated in the microampere range even at 10^{-10} torr. Bloomer and Cox[25] have shown that when oxygen is introduced at 10^{-6} torr background pressure, the emission current drops due to

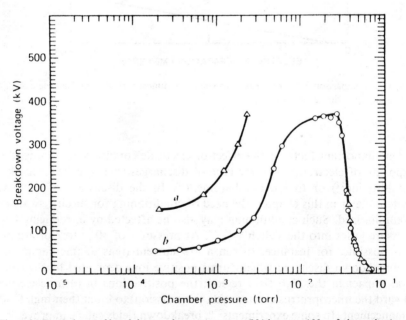

Fig. 4-2. Dependence of breakdown voltage on pressure (N_2) between 100 cm² electrodes (made of stainless steel, type 304) separated by 2.5 cm. Curve a is for partially conditioned electrodes. Curve b is for unconditioned electrodes. Reproduced from Ref. 12. Copyright © 1980 by John Wiley & Sons. Reprinted by permission of John Wiley & Sons, Inc.

a postulated increase in the work function of the electrode material. Other studies of similar effects have been summarized by G. A. Farrall in Ref. 12.

Finally, another distinct breakdown mechanism, already mentioned, has been identified. A microparticle on the electrode can contribute to breakdown due to its charge-induced motion across the gap or due to vaporization. Lafferty[12] and Latham[11], for example, discuss a variety of models for such breakdown that have been proposed in the literature on vacuum dc breakdown.

Vacuum and low pressure switches can be used in high power systems under "dc" conditions, as when a capacitor bank is charged over a period of seconds. Such switches are also used under transient conditions where voltage is applied for periods of microseconds or even less. Thus, the differences in breakdown behavior of low pressure insulation must also be examined under conditions of short pulses. The factor differentiating the behavior of the dc and pulsed breakdown is the time allowed for various mechanisms to operate relative to the delay time between voltage application to the electrode and the onset of breakdown. One of the initiation mechanisms which becomes unimportant in short-pulse breakdown is that of microparticle formation, because transit time in the gap may be longer than the applied pulse. Outgassing due to microcurrents could be also insignificant in determining pulsed breakdown. Other mechanisms which may determine the pulsed breakdown require, nevertheless, certain short amounts of time to establish the discharge, leading to a delay between the application of the fields at breakdown level and actual onset of the breakdown. The dominant breakdown mechanism in short-pulse regime is the "explosive" phenomenon involving thermal response of the emitter (i.e., heating and vaporization of the surface whiskers, mentioned above). Large number of studies, using very high temporal resolution (nanosecond) diagnostic techniques, have been performed to investigate this phenomenon which in its final phase creates cathode surface plasmas which move at $1-3 \times 10^6$ cm/s. A most detailed review of the phenomenon and some of the techniques used in such studies is given in Ref. 14. A less detailed summary is given in Chapter 5 of Ref. 11 and in Ref. 26.

While most studies have been conducted with small gaps and relatively low voltages, the experiment of G. B. Frazier[27] was performed using large (23 cm diameter) electrodes at high (500 kV) voltages to study the emission of electrons, which subsequently leads to gap closure by plasmas formed as a result of such emission. In this experiment, freshly prepared metal surfaces withstood ≥ 300 kV/cm for ≥ 40 ns without significant emission. Once the emission of electron current was initiated, the gap closure of given separation could be estimated from the cathode plasma expansion. The two-step breakdown process—initiation of the electron emission at current densities prescribed by the Child-Langmuire relationship[26] followed by plasma formation and closure of the gap—typically leads to delay time between the voltage application and discharge in the range of 10 μs, depending on the electrode surface preparation.

Evans et al.[28] have studied methods of enhancing the breakdown strength of vacuum gaps. The dependence of the breakdown field on the time to breakdown, such as would occur under pulsed conditions is shown in Fig. 4-3. The collection of data for several conductors, denoted in the figure, is for high vacuum and for gap separation ranging from fractions of millimeters to 15 cm and for applied voltages ranging from 20 kV to 2.0 MV.

Triggered Breakdown.   As in the case of gas spark gaps, it is possible to trigger breakdown of the interelectrode gap before self-breakdown takes place. Breakdown triggering can be achieved by injection of neutral gas or plasma, as well as by the use of a field-distorting third electrode.

Opening a valve to an outside gas reservoir allows the pressure in the gap to increase, at a rate determined by the gas flow dynamics, until a Paschen value for breakdown is reached. Because the thermal velocity of a gas is low, the rate of pressure rise limits the triggering speed.

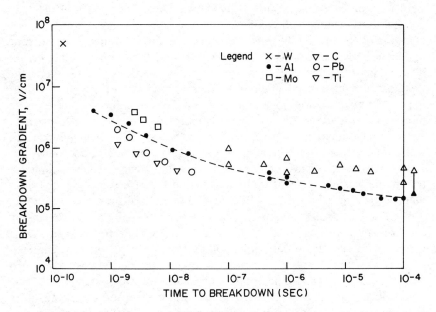

Fig. 4-3. Dependence of the breakdown field on the time to breakdown for electrodes made of metals indicated in the legend. Reproduced from Ref. 28.

Plasma injection into the gap region is achieved by producing an arc (or surface flash) between two electrodes, specifically introduced into the main gap region for this purpose, or between a separate trigger (igniter) and one of the main electrodes. Alternately, a pulsed laser beam can be injected to vaporize and ionize a small electrode surface area (requiring only a few millijoules of energy[29]) in a manner similar to that discussed in Chapter 3. When the metal plasma bridges the gap between the electrodes, an arc will develop between the electrodes, provided sufficient voltage appears across the arc to overcome its resistive drop. Plasma formation can be very rapid (using an appropriate trigger circuit) and its higher thermal speed provides faster breakdown than one obtained using injection of neutral gas.

Similarly, triggering can be obtained by using the spark over an insulating surface impregnated with materials such as titanium, which generates a rapidly expanding plasma very rich in hydrogen ions, as, for example, discussed in Chapter 3 of Ref. 12. Surface flashover triggering, generally, is found to be more reliable than breakdown of small triggering gaps[30]. (A surface flashover mechanism has been applied to switching of high pressure gaps, as well as to controlling the discharge distribution in laser cavities. A thorough discussion of such discharges in atmospheric and higher pressure regimes can be found in Ref. 31.) The construction of such a triggering device is shown in Fig. 4-4 reproduced from Ref. 12. The trigger is operated by applying a positive voltage pulse to the trigger lead. The ceramic tube wall flashes over, establishing an arc between the titanium hydride electrodes and releasing the hydrogen (and titanium vapor), which ionizes and enters the main electrode gap. As the plasma spreads out in the gap, a high current discharge is established. With a trigger energy of $> 10^{-2}$ J, very rapid switching of the main electrodes is achieved. Using a 10 A trigger current, the main electrode discharge starts in less than 100 μs (with a jitter time of about 30 ns) when 30 kV is applied across the main electrodes[12].

Another version of the flashover technique has been reported in Ref. 30. It utilizes a current pulse to vaporize part of a conducting film on an insulating surface located between the cathode and the igniter electrode. The subsequent switching arc regenerates the film by deposition of the electrode material on the trigger surface. Obviously, for efficient ignition of breakdown (for all methods of plasma formation), the plasma must be permitted to bridge the gap effectively. This necessitates the location of the trigger plasma source in a manner that assures an unobstructed flow of plasma into the gap.

The use of a field-distorting electrode to trigger vacuum gaps involves an operation similar to that discussed in the preceeding chapter. The trigger pulse leads to a large electric field, exceeding the breakdown strength of one part of the switch. Its breakdown then leads to the breakdown of the

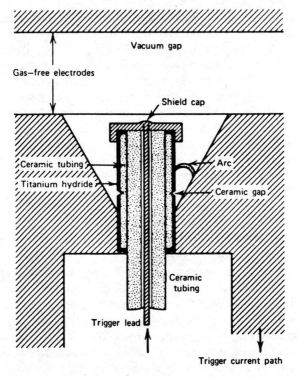

Fig. 4-4. Construction details of the triggering device for controlling breakdown time in a vacuum gap switch. Reproduced from Ref. 12. Copyright © 1980 by John Wiley & Sons, Inc. Reprinted by permission of John Wiley and Sons, Inc.

remaining part of the main gap. The triggering speed should thus be similar to that of the self-breakdown gaps charged to the voltage determined by the trigger circuit voltage and risetime. (See the end of Chapter 3 in Ref. 12 for more material on triggering).

Surface Flashover

Because surface discharge phenomena have been applied not only to switching but also to white light sources[33] and as means of powering gas laser cavities[28], these discharges have been studied over a broad range of pressures, from vacuum to above atmospheric. Many investigations of vacuum discharges were also instigated by attempts to reduce the inductance and increase the reliability of intense relativistic electron beam diodes[26, 32] i.e., to increase the hold-off voltage of the vacuum-insulator interfaces forming the diode structure. Surface discharges are governed by breakdown criteria analogous to that in gases, liquids, and vacuums. Various types of surface-electrode configurations are used, shown in Fig. 4-5. The top configuration, assuming that no breakdown occurs through the insulator (with a dielectric constant, ϵ), leads to long discharges along the surface until the energy feeding the discharge is exhausted or until a ground point of the "substrate" electrode is reached by the streamer. The middle configuration establishes the discharge along or near the surface. The distance, Δ, separating the electrodes from the surfaces can be finite, or the electrodes can be in contact with the surface. The bottom configuration uses the dielectric in yet another manner. These three main configurations can be outfitted with additional electrodes or subjected to ionizing radiation to trigger the discharge.

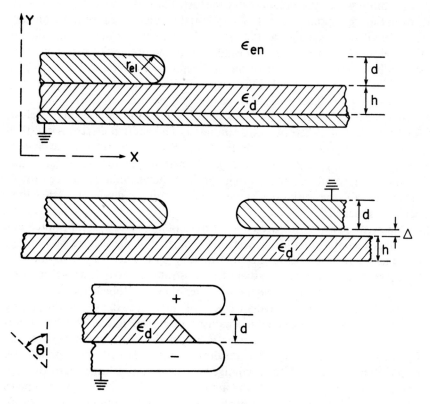

Fig. 4-5. Types of electrode-dielectric surface configurations used in discharge streamer studies and in surface flashover switches.

Untriggered Initiation.   The geometries associated with surface discharges shown in Fig. 4-4 do not lend themselves to a simple investigation of breakdown mechanisms because of the complication of determining the electric fields due to the presence of the insulator with a dielectric constant, ϵ_d, which sometimes is much higher than that of a vacuum. To alleviate this situation, P. N. Dashuk et al.[33] analyze the features of the surface discharge by considering electric field distributions for the first (top) case shown in Fig. 4-5. In this analysis, they compute the E_x and E_y components of the electric field in the region of the high voltage electrode and the interface. The calculations range over the ratio $d/h = 0.01$ to 50 for various radii of curvature of the electrode, r_{el}, such that $r_{el}/h = 0$ to 0.5, and for $\epsilon_d/\epsilon_{en} = 1$ to 100; ϵ_{en} refers to the surrounding environment. The results given graphically in Ref. 33 show that for distances from the interface greater than the electrode radius $E_x > E_y$ for $d/h = 0.5$ to 1.0. The x-component determines the process by which the discharge is initiated. Along the surface, E_x reaches a maximum where the breakdown can start. For distances $d/h = 0.3$ along x from the triple point, $E_y > > E_x$, essentially independent of the geometry. Charges of the same sign as the high voltage electrode may accumulate in this region if time for migration is allowed in the prebreakdown stage, substantially weakening E_y. Increasing ϵ_d/ϵ_{en} leads to an increase in E_y and E_x in the region $d/h > 0.2$ to 0.7, which explains the experimental dependence of the breakdown voltage on ϵ_d/ϵ_{en} (Ref. 1).

In a vacuum, when electric fields of 10^4 V/m at the cathode are reached, field emission of electrons can be initiated. The emitted electrons can then be accelerated (in the vacuum) in the fields distorted by the surface charges and accelerated to ≥ 100 eV, which is sufficient to generate secondary emission, leading to an avalanche discharge[33]. If the vacuum is replaced by atmospheric air, the initial discharge can be a volume discharge rather than being channelled[34].

The breakdown of surfaces (including ceramic, cement, and printed circuit board samples) with the electrode geometry of the second case in Fig. 4-5 was studied at pressures of 1-5 atm by S. T. Pai et al.[35], specifically, to determine the effect of the dielectric substrate on breakdown voltage and on channel formation. In the model which they developed, the basic effect of the dielectric surface is enhancement of the charge density in the gap by the electric field enhancement mechanisms discussed in Ref. 33, which holds generally for all pressures. Their basic model yields the analytic expression for the breakdown voltage. The results compare well with the experimental results, where the breakdown voltage is measured as a function of the separation Δ given in Fig. 4-5. In the model, production of electrons competes with electron loss or diffusion (including electron trapping by the surface). The breakdown voltage for an electrode-surface separation Δ is

$$V_B = A \left[\frac{G}{\Delta^2} J_0 \left(\frac{2.4\,r}{\Delta} \right) \exp \left(\frac{t-t_0}{\tau} \right) + n_0 \right]^{-1}, \quad (4\text{-}3)$$

where A, G and n are constants determined experimentally, J_0 is the zero-order Bessel function, r is the spatial coordinate, R is the distance between the surface and electrodes, t is the time after the field is applied across the surface, t_0 is charge saturation (to a level n_0) time and τ is a time constant related to net rate of production of electrons and to the diffusion coefficient. This relationship shows that as Δ increases from zero, V_B decreases to a minimum and then increases assymptotically to the value near that for $\Delta = 0$. The general effect of the dielectric surface is to lower the gap breakdown voltage.

To provide a practical guide for a breakdown in a vacuum along a surface, I. Smith[36, 37], noting that preventing the electrons emitted from the electrodes from reaching the insulator surface (and, consequently, forming secondaries) should increase the breakdown strength of the dielectric-vacuum interface, investigated the surface-electrode geometries shown at the bottom of Fig. 4-5. Figure 4-6 represents the dependence of the breakdown field E_B on the angle, θ, from the normal to the electrode surface and the inclination of the surface as indicated in Fig. 4-5. The data of Fig. 4-6 is for epoxy resin. Reference 36 also shows that other materials (polyethylene, Lucite, and glass) have similar dependence on the inclination angle, θ. With the

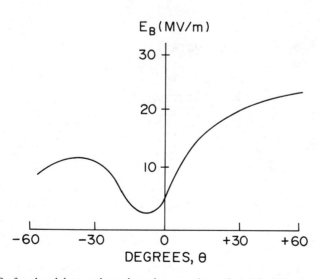

Fig. 4-6. Surface breakdown voltage dependence on the angle θ defined in Fig. 4-5. Reproduced from Ref. 37. © 1982 IEEE.

exception of the poor hold-off strength of glass, the other materials have values of breakdown strength similar to those of the epoxy resin.

These data have been taken for voltages applied across insulators in short pulses, 30 ns and 1 μs in duration. Noting that flashover is a statistical process, so that the larger the insulator, the more likely it is to break down at a given value of the electric field, J. C. Martin has analyzed such data from this viewpoint and presented it in an unpublished form[38]. His result is quoted in Ref. 26 (with units of the time of voltage pulse given mistakenly in nanoseconds, rather than in microseconds).

Figure 4-7 is the potential plot for the insulating truncated conical post between parallel plate electrodes of the type used in tests establishing the breakdown dependence on the angle (seen in Fig. 4-6). It is clear from such plots that the effect of inclining the wall (with θ being defined as positive) is to prevent electrons emitted from electrodes from encountering the insulator and hence from multiplying. If the polarity is inverted, it is clear that the process described is assisted and can occur anywhere on the surface. Thus, the shape of the curves at small angles is now explained; and at larger negative angles the breakdown strength increases again because electrons emitted from the insulator tend to meet the insulator again before gaining much energy in the accelerating field. Thus, in case of a positive cone (e.g., Lucite with $\theta = 45°$) much greater fields can be withstood than on cylinders.

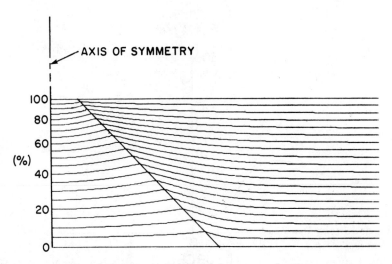

Fig. 4-7. Distribution of equipotentials established by a voltage applied to parallel plane electrodes in a vacuum, where there is a conical post between the electrodes. Equipotentials are given as a percentage of the applied potential.

Nevertheless, a similar breakdown occurs in overall fields of 200 kv/cm. In a small region near the anode, gradients of 5×10^5 V/cm exist because of the polarization of the dielectric, and it is possible that field emission from the insulator could reach current levels of the magnitude required to begin discharging the surface there. Such emission would certainly be non-uniform with respect to distance from the anode and would depend on local irregularities. It could produce regions with larger fields, as well as tend to enhance the electric field to become more nearly parallel to the surface so that multiplication can take place. The degree of field enhancement due to polarization and the area it affects grow with increasing cone angle. This is why still further change in the cone angle yields poorer performance; it accounts, as well, for the poorer properties of glass and pyrophylite, whose dielectric constants are roughly twice those of the plastics. (Glass also has a high secondary electron coefficient.)

When diodes for the production of 10^{12} watt electron beams were being developed, J. D. Shipman[39] of the Naval Research Laboratory and P. Spence[40] of Physics International, Inc., reviewed and independently developed scaling relations for large-area insulators between a set of electrodes. Their unpublished work is summarized by J. P. Van Devender and correlated with J. C. Martin's scaling for the breakdown field, E (in mks units)[32]:

$$Et^{1/6}A^{1/10} = 7 \times 10^5 \qquad (4\text{-}4)$$

Here A is the area of the insulator-vacuum interface and t is the time during which A is stressed to more then 89% of its maximum value. E is defined as the ratio of the breakdown voltage to the distance between the electrodes. In the formulation by Shipman for the optimum design of low inductance diodes, the electric field is within 20° of the normal to the surface and is uniform across the insulator except at the anode and the cathode:

$$E_s t^{1/6}A^{1/8.5} \leq 5 \times 10^5 \qquad (4\text{-}5a)$$

$$E_T t^{1/6}A^{1/8} \leq 1.6 \times 10^6 \qquad (4\text{-}5b)$$

$$E_{el} < 5 \times 10^6 \qquad (4\text{-}5c)$$

where E_s and E_T are respectively the greatest electric field parallel to the insulator surface and the greatest value of the total field, and E_{el} is the field parallel to the surface near the cathode or anode intersections with the insulator. Expressions (4-5) retain the same time dependence as Eq. 4-4, the area dependence is somewhat more conservative, based on the observations of large interface behavior[39].

The approach by Spence[40] to determining the optimum insulator interface with a vacuum is different in that his recommendation for improved hold-off is to increase the electric stress in direct proportion to the increasing distance from the cathode. His criteria apply to insulators with the electric field pointing from the vacuum region into the insulator. Spence retains the area dependence of Eq. 4-4:

$$E_s t^{1/6} A^{1/10} \leq 5.6 \times 10^5 \qquad (4\text{-}6a)$$

$$E_T t^{1/6} A^{1/10} \leq 1.8 \times 10^6 \qquad (4\text{-}6b)$$

$$E_{el} < 4.5 \times 10^6 \qquad (4\text{-}6c)$$

When, however, the electric field points out of the insulator, the criteria are

$$E_s t^{1/6} A^{1/10} \leq 3.7 \times 10^5 \qquad (4\text{-}7a)$$

$$E_T t^{1/6} A^{1/10} \leq 5.4 \times 10^5 \qquad (4\text{-}7b)$$

$$E_{el} \text{ is undetermined} \qquad (4\text{-}7c)$$

These formulas with the aid of equipotential solvers such as FFEARS[41] or JASON[42] or modeling with resistive paper or electrolytic tanks can be adapted easily to the design of vacuum switches to optimize surface hold-off, for example, for minimum switch inductance. Two versions of a six stage insulator separating water dielectrics from vacuums are shown in Fig. 4-8, where shaping of the dielectrically floating electrodes is utilized to maintain uniform electric field or potential distribution on all insulator stages so that the degraded performance due to the dominance of the weakest stage is eliminated. A vacuum envelope of the type shown in Fig. 4-8 can be used for a very high voltage vacuum switch or, as it was actually used, for an electron beam diode in series with the flashover prepulse switch (discussed in Ref. 26).

Triggered Initiation. In developing the surface flashover criteria discussed in the preceeding section, care was taken to insure that no ultraviolet radiation was incident on the electrically stressed surfaces. Shipman[39] cautions that small incidental sources of ultraviolet radiation, such as sparking at joints of metallic surfaces resulting from high current flow, if allowed to irradiate the surface, lower the insulator surface breakdown strength. Generally, radiation incident on the dielectric surfaces between the electrodes leads to premature breakdown if applied early during the increasing voltage pulse, or initiates breakdown when irradiation occurs after the volt-

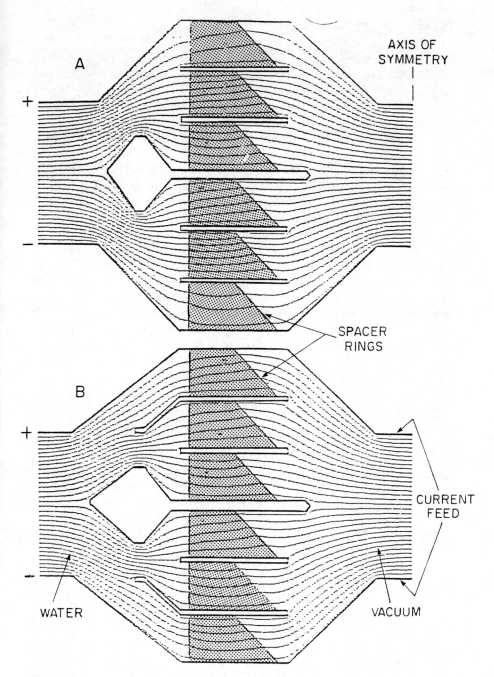

Fig. 4–8. Equipotential distributions in two water dielectric-to-vacuum interfaces using metal grading rings between lucite rings separating the current feed. Such housing designs are used for integrating the pulse line with the vacuum switch or with electron beam diode. Reproduced from Ref. 32.

77

age has been applied across the surface. Ultraviolet radiation plays a similar role in initiating surface discharge, for example, when the electrodes of the vacuum gap are irradiated. In both cases, the photon energy is sufficient to free electrons from the surfaces, allowing them to be accelerated in the applied electric field.

A detailed study of the effects of UV radiation on the flashover in a vacuum (2 to 4 × 10 $^{-5}$ torr) is reported by C. L. Enole et al.[43] for the dielectric surface geometry shown in Fig. 4–5 (bottom) for angles $\theta = +45°$ and $-45°$ and for $\theta = 0°$. Polyethelene, polystyrene, and Teflon insulators were studied at electric field stresses ranging from 10 to 70 kV/cm. The UV irradiation contained photons with energies up to 12 eV. (All insulators were unaffected by photons with energy of less then 3 eV, regardless of power). The difference between positive (actually shown in Fig. 4–5) and negative geometry (where the electrons are accelerated into the insulator surface) was quite dramatic, showing that surface breakdown under UV illumination tends to occur at lower field levels for the surface-electrode configuration, opposite to the case without irradiation. These results are in agreement with those A. A. Avdienko[44] obtained for different time and spectral regimes. The proposed model in Ref. 43 explains such behavior on the basis of enhanced positive surface-charge (and the accompanying increase in the electrical field tangential to the surface), when electrons do not re-enter the insulator surface (positive θ), as well as the possibility of release of desorbed gases which create a path for a current flow over a large cross-sectional area.

Less suitable for switch applications is the formation of light-initiated discharges at pressures higher then those used by Enloe et al.[43], which also has been studied recently. For example, A. G. Bedrin et al.[45, 46] have examined discharges in the pressure regime from 0.25 to 1.0 atm (air). Their irradiation pulses are several hundred microseconds in duration and voltage and current waveforms are a few tens of microseconds in duration. In that regime, no discharge initiation occurs for threshold irradiance (from a Xenon flashlamp) below 1 to 3 J/cm^2 and for very low fields (100–300 V/cm) across the gap. The light initiated surface discharge occurs when a vaporized layer forms near the surface of the insulator (and expands at a velocity of 10–100 m/s), leading to a substantial (up to a millisecond) delay between the initiation of the discharge and the onset of high current. Because the time delay is independent of the gas used, the authors conclude that the main discharge mechanism operating at low electric fields (10 to 100 times lower than breakdown fields without irradiation) is a hot gas layer formed by the vaporization of the insulator (ablating at a rate of 1 g/cm^2 s). Thermal ionization by the small (10 to 20 A) prebreakdown current leads to high current discharge. The discharge itself, at least initially, is a volume discharge. Initiation of discharges on dielectric surfaces at atmospheric and higher pressures is discussed in a very complete review by W. J. Sarjeant[31].

The use of a trigger (field-distortion) electrode to initiate surface discharges has been developed by I. Smith for very precise high voltage switching applications[37]. Such a third electrode triggers the surface discharges by changing very rapidly (and in a controlled manner) the distribution of the electric field in the main gap of the switch, analogously to that in gas, solid and liquid switches described in the Chapters 3, 5, and 7, respectively.

SWITCH DESIGNS

Gap Discharges

Vacuum gap switch features that exploit the insulation properties of vacuum during hold-off and of vacuum conduction mechanisms during switch discharge have now been described. There are other features of vacuum gaps that are common with gap switches using gaseous, liquid, and solid dielectrics in the interelectrode region. These can be recognized by comparing designs found, respectively, in Chapters 3, 5, and 6. For example, while the micro-structure of the electrode surface is very important in determining vacuum gap breakdown, the gross geometric enhancement of the electric field is an effect to be considered in all switch designs independent of the insulating medium. Helpful hints relating to the electrode design, as well as quantitative graphs of the field enhancement can be found in Ref. 12, Figs. 2.8, and 2.9, reproduced in this work as Fig. 4–9. It shows the extent of field enhancement for geometries commonly used in the switch designs and supplements similar scalings discussed in Chapter 3, e.g., Fig. 3–3. Another design point common to all gap switches is the treatment of the "triple point," where the electrode surface meets the gap insulator-dielectric wall interface (e.g., as discussed in Ch. 6 in conjunction with diaphragms separating the dielectrics) and of switch connections to optimize the transfer of the energy from the source to the load, as illustrated in Fig. 4–10. The latter point is treated fully by Ware et al.[47] in discussion of their triggered switch, considered later in this chapter.

Breakdown Gaps. The simplest vacuum switch is a cylindrical chamber of insulating material with flat or rounded electrodes at each end. For high voltages above 50 kV, the best approach is to stack sections of about 50 kV each for reasons suggested in the "Surface Flashover" subsection above. Additionally, the use of sectioned switches results in a lower prefire rate because a breakdown in one section can be prevented from propagating to other sections by a complete screening of the resulting UV radiation from

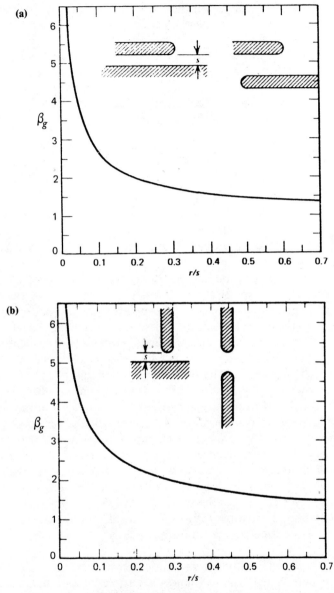

Fig. 4–9. (a) Electric field enhancement factor shown as a function of the aspect ratio r/s, where r is the radius of the rounded plate parallel to a plane. (b) Electric field enhancement factor at a point of closest approach shown as a function of the aspect ratio r/s for a rounded edge plate perpendicular to a plane. Reproduced from Ref. 12. Copyright © 1980 by John Wiley & Sons, Inc. Reprinted by permission of John Wiley & Sons, Inc.

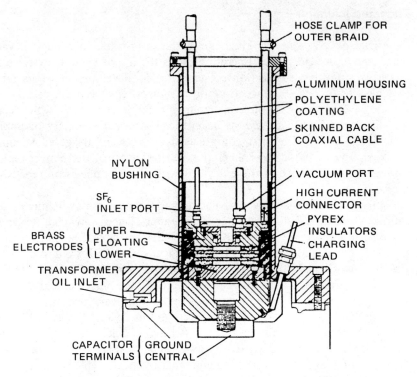

Fig. 4–10. Cross-sectional view of the graded vacuum spark gap shown with electrodes designed for cable connection to the capacitor bank. Reproduced from Ref. 47.

illuminating electrodes and surfaces of other sections. This technique is further discussed in conjunction with triggered vacuum gaps below.

To eliminate the effects of field enhancement due to electrode and electrode-interface ("triple point") gross geometric effects, it is possible to shape electrodes so as to obtain an essentially uniform field over most of the surface. Farrall (Ref. 12) discusses such electrode designs, as mentioned earlier. Because the rule-of-thumb for field uniformity for one of these profiles, the Rogowskii profile, requires that the flat central section extend to a diameter at least five times greater than the electrode gap, the price for elimination of field enhancement is a large switch size.

The switching of large currents, of the order of 100 kA, deposits a large amount of energy in the walls and in the electrodes and leads to a substantial increase in the pressure in the switch chamber, so that the hold-off voltage cannot be maintained. To remedy this, constant pumping of the chamber is necesary.

Triggered Initiation

Conventional Switches. The large scatter in breakdown voltage of vacuum gaps in dc and pulsed operation limits the use of vacuum systems to relatively primitive applications. Therefore the development of vacuum switches has tended to concentrate on designs of triggered devices. These include development of commercial units[48, 49] and devices which, while being used in specific systems, are sufficiently versatile for a variety of applications[2, 47, 50, 51]. The first comprehensive review of triggered vacuum spark gaps was provided by Komelkov[1]. Burkes[3] has provided a more recent survey of the literature of these devices, with some emphasis on repetitive operation.

Figure 4–11 shows a triggered vacuum gap intended for switching applications requiring a wide operating voltage range. The gap switch is designed to be used with energy storage of up to 30 kJ and a charge transfer of 2 C[48]. The operating range (when a 40 A trigger pulse of 12 kV into an open circuit is used), shown in Table 4–1, allows this switch to be used also for crowbar operation. In the last case, a high voltage must be held-off; the switching action must occur at low voltage and at peak current.

A high current (400 kA) and low jitter (± 4 ns) switch designed for low inductance capacitive storage circuits (with the switch inductance being about 5 nH) has been designed by Ware et al.[47]. Fig. 4–10 is the cross-sectional view of the sectioned vacuum spark gap, as depicted in Ref. 47. Its mechanical design provides long lifetime (of about 1000 discharges of 10 kJ storage capacitors). The design incorporates a stack of Pyrex insulator discs (6.3 mm thick, 86 mm o.d., 32 cm i.d.) and two electrically floating brass electrodes (76 mm o.d. and 13 to 26 mm i.d.). The central hole in the electrodes allows the radiation from the trigger spark plug to reach all switch electrodes. The insulator diameter is sufficiently large to keep the discharge away from the walls. The insulator walls are inclined 20° to 40° with respect to the axis to increase the voltage hold-off suggested by the breakdown voltage optimization of Fig. 4–6. The outside electrodes compress the metallic and insulating rings by threading them into a cylindrical nylon bushing, as seen in Fig. 4–10.

Reference 47 also provides details of a low inductance connection between the switch terminals, the storage capacitor collector plate, and the coaxial cable feeding the current to the load. The scheme requires pressurization with SF_6 gas.

Ref. 47 also provides the description of the trigger plug, consisting of a 2 mm diameter tungsten rod inside a 4.75 mm diameter alumina jacket.

Fig. 4–11. Triggered vacuum spark gap switch made by EG&G. Reproduced from Ref. 48.

Mechanical Dimensions

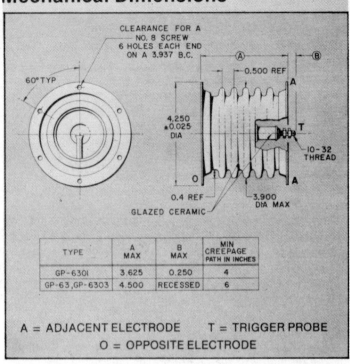

CLEARANCE FOR A
NO. 8 SCREW
6 HOLES EACH END
ON A 3.937 B.C.

60° TYP

Ⓐ Ⓑ

0.500 REF

4.250 ±0.025 DIA

A

T

10-32 THREAD

A

0.4 REF

O

3.900 DIA MAX

GLAZED CERAMIC

TYPE	A MAX	B MAX	MIN CREEPAGE PATH IN INCHES
GP-6301	3.625	0.250	4
GP-63, GP-6303	4.500	RECESSED	6

A = ADJACENT ELECTRODE   T = TRIGGER PROBE

O = OPPOSITE ELECTRODE

Table 4-1. Production Model
Switch Characteristics.[48]

EGG Switch Model No.	Self-breakdown (Minimum)	Operating Voltage	
		(Min)	(Max)
GP63	65 kV	0.3 kV	50 kV
GP6301	40 kV	0.3 kV	40 kV
GP6303	120 kV	1.0 kV	100 kV

The jacket is fitted into a brass housing and nylon bushing and includes an O-ring seal to make it vacuum-tight.

A newly assembled switch must be conditioned by low current discharges. During conditioning, the outgassing is quite significant, reaching pressures of 0.1 torr after each discharge. As the outgassing decreases to a negligible level (after 10 to 15 discharges), the switch becomes ready for operation at a voltage level of 50 kV. Back-filling of the switch to 5–25 mtorr of krypton was found to provide the fastest triggering (~ 15 ns) while maintaining good hold-off.

Various practical points, such as a need to avoid an evacuated line connection between the switch and vacuum pump (because the high voltage electrode can discharge to the pump) are also discussed.

Higher current operation, at up to the 2 MA level, can be achieved with the switch design developed by Aretov et al.[4] This versatile switch is designed to hold-off dc voltages of 30 to 50 kV, similar to the switch of Ref. 47, by adjusting the number of standardized sections (41 cm in diameter) for required performance. A single device can switch 100 kJ for a small number of discharges and 50 kJ for thousands of discharges. The authors recognize two separate pressure regimes that simultaneously affect the switching delay (relative to the trigger) and the hold-off voltage. For pressures below 10^{-3} torr, delay times of many microseconds were observed when a trigger spark was used to provide UV radiation of all switch sections (through appropriately positioned holes in each electrode) as well as to inject plasma into at least the first switch section. When the pressure was in a 10^{-3} to 10^{-1} torr range, much faster (fraction of a microsecond) delay was obtained. Breakdown dependence on pressure was found to depend strongly on the electrode separation of the switch sections. Switch performance over a broad range of parameters, as well as a methodical discussion of the construction, is summarized by the authors in Ref. 4; this summary can serve in the design of vacuum switches for which very stable operation in low inductance circuits is required.

Vacuum switches have also been developed specifically for crowbar applications[50, 51] at current levels below 100 kA, with charge transfer via the switch of up to 100 C. Reference 50 shows that a surface flashover trigger unit similar to that already described provides discharge delay and jitter time similar to that in Ref. 4. Ref. 51 discusses triggering of 1 cm gaps, which can initiate the discharge with gap voltages as low as 50 V.

Similar switches have been studied by Milde et al.[2] Milde indicates, without further description, that relatively high power vacuum switches can be repetitively pulsed, at least for short bursts. He finds that a four-section switch recovers to a hold-off voltage level of 80 kV in 10 μs after passing a current of 20 kA. (At lower operating voltages, 2.5 kV, 9 kV and 50 kV, continuous repetitive operation for many hours was obtained at frequencies of 35, 10, and 1 kHz respectively[2].)

Current Density-Controlled Discharges:

Pseudo-Spark Switches. The vacuum gap switches considered so far have not been designed to control the discharge plasma distribution. Clearly, reproducible control of the current density on the electrodes can be advantageous to switch reliability and lifetime. By keeping the current density at the electrodes low, little metal vapor and surface contaminants will be released during switching. This can improve switch recovery and preserve the insulator walls from deterioration. Christiansen and Shultheiss[52] invented a method for controlling the density of the discharge current using plasma triggers in a cavity formed by one of the electrodes. (A similar scheme was later used for establishing a conducting plasma in the plasma erosion opening switches described in Chapter 9). They chose to call such discharge systems "pseudo-spark" discharges. However, this led to a very reliable low pressure switch, with the latest version described by Boggash and Riege[53].

Figure 4-12[53], above, shows schematically the current distribution between electrodes with cavities behind the juxtaposing electrode surfaces. Below is the cross-section of a switch designed to operate at 400 kA (in parallel modules) with a 20 kV hold-off. Precise triggering of the switch has been developed to allow parallel operation of at least four or more switches[53]. The gas inlet also accommodates the plug-in trigger. Two types of low energy triggers can be used to obtain approximately 100 ns delay and ± 10 ns jitter; the latter can be as low as ± 1 ns for some operational parameters. One of the triggers is the surface flashover switch of the type described above. The working gas is helium at a pressure of 0.01 to 0.05 mb.

The use of cavities not only allows the current distribution to be controlled, but also serves to confine the electrode debris away from the interelectrode space, where electric fields appear before switching initiation. This suggests that this switch design is better adaptable to repetitive operation than are the conventional vacuum gaps. In fact, such switches have been

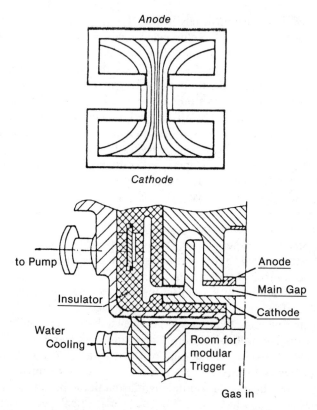

Fig. 4-12. Above, an idealized pseudo-spark switch and the distribution of current in such gaps. Below, a high current switch developed by the authors in Refs. 5 and 53 is shown.

© 1985 IEEE.

operated for a medium power range, switching 20 kV at up to 10 kA with a repetition rate of 100 kHz[54].

More detailed characterization of the pseudo-spark switch is discussed by D. Bloess et al.[55] An important observation is that the switch internal resistance (of the conducting plasma) driven by 50 Ω cables drops to less than 0.1 Ω within a few nanoseconds after the breakdown. No saturation (or "erosion") of the internal resistance was noted for 4 μs. This contrasts with the behavior of similar plasma discharges, *at higher currents* in the plasma erosion opening switches described in Chapter 9. The switch operating at 30 kV maintained ±2 ns jitter time after 2×10^6 discharges with pulse duration of 1–2 μs. The authors[55] further suggest that a multi-cavity switch could be developed to operate at voltages higher then 100 kV.

Ionization Initiation. A different method of controlling the current distribution of the low pressure switch was investigated by G. Lauer et al.[56] With a preinjected low pressure (0.1 mb) gas in the switch consisting of electrodes with relatively large surfaces, spark triggers distributed over the surface of one electrode were used to ionize the gas leading to stable volume conduction for short duration (10 μs) discharge pulses. Because of the gas injection into the inter-electrode region, the pressure during the application of voltage across the switch was much higher in the gap than on the dielectric wall surfaces. This provided for limiting the current to the gap region where ionization by the trigger caused a volume discharge. This switch, described in Ref. 56, operated at up to 250 kV and 40 kA. The current density of approximately 100 A/cm^2 assured fast recovery of the switch, leading to full recovery after 100–1000 μs, depending on gas injection rate. These recovery rates suggest that this type of a switch is a candidate for development as a high power switch operating repetitively at rates of at least 1 kHz.

Surface Flashover

Because of the broad range of application of vacuum and low pressure switches, it cannot be claimed that a single design prescription exists for constructing surface discharge switches. Thus it is best to survey a variety of switches that have been successfully developed and use those as a guide for any specific application. A variety of such surface flashover switches have been developed for applications ranging from switching tens of kiloamperes (with potential to operate repetitively at 1–10 kHz[37]) to tens of megamperes[57], with hold-off voltages up to 250 kV[37]. Their operating pressures vary from atmospheric[57] to low vacuum[34].

Reinovsky et al.[57] have developed a self-break switch for operating at a level of tens of megamperes. The switch is a rather large device, carrying 5–10 MA/m; it is configured in a hexagonal shape connecting six rows of capacitor banks storing a total of 9 MJ when charged to 120 kV. Its design resembles the configuration shown in the middle of Fig. 4–5, with a return current strap underneath the insulator. The return strap is a key to assuring that surface discharge is initiated with a large number of parallel channels. The discharge surface is opened to air. With 3 μs charging (to 100 kV), about 100 parallel discharge channels develop over the entire 3 meter length of the switch, carrying 14 MA.

An intermediate-current switch, also operating in air, has been described by Grigoriev et al.[58] A multichannel discharge along a surface (with electrode spacing of 16 cm) is used in switching a capacitor bank with voltage ranging from 0 to 40 kV. A field-distortion electrode is used for triggering. The 10^{12} V/s trigger pulse results in 40 ns delay with a jitter of ± 10 ns. The switch resistance depends on current and varies from 5×10^{-2} Ω at 100 kA to less than 10^{-2} Ω at 500 kA. The linear current density obtained in the configuration used by Grigoriev and his collaborators is 5MA/m, the same as that of Reinovsky et al.[57], as noted above. If three other switches (reported in Introduction section of Ref. 35) are added to those described in Refs. 37 and 57, the five varieties of switches cover all the significant design issues of surface flashover switches.

The special design feature incorporated in the low-current switch of Ref. 37, which optimizes the voltage hold-off, is the angled vacuum dielectric interface surface such as that shown at the bottom of Fig. 4–5. The electric field at flashover across the interface is highly reproducible, independent of the electrode spacing and only weakly dependent on the voltage duration in the 10 ns to 10 μs range. Both an inside surface of a hollow cone and outside surface of a solid cone are used, with magnetic forces of the discharge pushing the discharge away from, and onto, the surface respectively. Trigger electrodes for field distortion initiation of the discharge are used either in a vacuum or imbedded in the material (Lucite and Delrin) of the cone. The latter configuration, when triggered with a negative potential, produces sub-nanosecond jitter over a voltage range of 3:1. Substituting alumina for the plastic increases the jitter time to tens of nanoseconds. In recovery tests of the switch, Smith[37] has found that much more rapid recovery occurs when the plastics are replaced by alumina. Recovery to 80 kV requires only 100 μs (compared to about 800 μs for the plastic) and to about 200 kV in 200 μs.

Other references (35, 45, 59, 60) provide helpful guides for specialized surface flashover switch design requirements. Ref. 45 (and Ref. 43) describes the effects of ultraviolet radiation on electrically stressed surfaces. S. T. Pai et al.[35] and P. N. Dashuk et al.[59] deal with the erosion of the dielectric surface at pressures of 1–5 atm and 1 atm respectively. The latter work provides exhaustive data on the erosion rates of plastic surfaces under varying discharge conditions. Both references outline the model describing the discharges and the erosion rates of the discharges with currents of no more then a few kiloamperes at pressures in the millibar range.

CONCLUSION

The gap and surface flashover switches considered in this chapter provide alternative designs for high power switching. Low inductances, mechanical impulses that are weaker than in high pressure or solid dielectric switches, low jitter time, long lifetime, and potential for repetitive switching at high frequency are some of the criteria used in selection of such switches for specific high power applications. Pulser system design considerations may also influence the choice of a vacuum or low pressure switch for these applications. For example, inductive pulse power sources using opening switches described in later chapters of this book also employ closing switches; common vacuum chamber housing opening and closing switch elements may simplify the electrical and mechanical design of pulsers.

Some of the more prominent research on switching processes and the consequent development of vacuum and low pressure switches has been reviewed earlier in this chapter. Perhaps if more attention were directed at plasma decay processes in the gaps and on surfaces, a practical high power repetitive switch, operating at rates approaching 100 kHz, could be developed. Some of the issues that clearly require further development are related to controlling the switching plasmas: their timing, density distribution, and interaction with magnetic fields as well as methods for accelerating their decay. While plasma decay control via the use of magnetic pressure for blow-out from the electrode region and through interaction with the walls is a known method for enhancing switching performance, the deliberate control of plasma sources for switching applications has received attention only recently.

REFERENCES

1. The most extensive review of vacuum switch designs and their performance characteristics is given by (Ref. 4, Ch. 3) V. S. Komelkov, "Technology of Large Impulse Currents and Magnetic Fields", translated by Foreign Technology Division, Wright-Patterson AF Base, OH, Report No. FTD-MT-24-992-71, from *Tekhnika Bolshikh Impulsnykh Tokov i Magnitnykh Poley,* Atomizdat, Moscow (1970).
2. H. I. Milde, C. J. Shubert, and R. Harrison, "Repetitive High Power Switching Technology," *Proceedings of the Workshop on Switching Requirements and R&D for Fusion Reactors,* Electric Power Research Institute, Palo Alto, CA, Special Report ER-376-SR (1977).
3. T. R. Burkes, "A Critical Analysis and Assessment of High Power Switches," Naval Surface Weapons Center (Dahlgren, VA), Report NP 30/78 (1978) (same as Ref. 3, Ch. 3).
4. G. N. Aretov, V. I. Vasilev, M. I. Pergament, and S. S. Tserevitinov, "Electrical Strength of Vacuum Switches," *Sov. Phys. Tech. Phys.* **11**, 1548–1555 (1967). See also same authors, "Delay Characteristics of a Vacuum Disc Switches", *Sov. Phys. Techn. Phys.* **12**, 90–96 (1967).
5. E. Boggasch and H. Riege, "The Triggering of Pseudo-Spark Switches," *Proceedings of the XVII International Conference on Phenomena in Ionized Gases,* Vol. 2, Contributed Paper I-10, Budapest pp. 567–569 (1985).
6. A. M. Andrianov, V. F. Demichev, G. A. Eliseev, P. A. Levit, A. Yu. Sokolov, and A. K. Terentev, "Pulse Generator Producing a High Power Current," *Instrum. Exper. Tech.* **14**, 124–126 (1971).
7. D. R. Detman, R. Dollinger, W. J. Sarjeant, "Pulsed Power Characterization of Metal

Plasma Arc Switches (MPAS)," *Digest of Technical Papers of the Fourth IEEE Pulsed Power Conference,* Albuquerque, NM, IEEE Cat. No. 83CH1908-3, pp. 762–765 (1983).

8. W. R. Baker, "High Voltage Low Inductance Switch for Megampere Pulse Currents," *Rev. Sci. Inst.* **30**, 700–702, (1959).

9. W. D. Coolidge, *Am. J. Roentgenol. Radiat. Ther.* **19**, 313, (1928).

10. J. M. Meek and J. D. Craggs, *Electrical Breakdown of Gases,* John Wiley & Sons, New York (1978).

11. R. V. Latham, *High Voltage Insulation: The Physical Basis,* Academic Press, New York (1981).

12. J. M. Lafferty, ed., *Vacuum Arcs Theory and Applications,* John Wiley & Sons, New York, (1980).

13. E. A. Litvinov, G. A. Mesyats, D. I. Proskurovskii, "Field Emission and Explosive Electron Emission Processes in Vacuum Discharges," *Sov. Phys. Usp.* **26**, 138 (1983).

14. R. J. Noer, "Electron Field Emission from Broad Area Electrodes," *Appl. Phys. A* **28**, 1 (1982).

15. D. Hinshelwood, "Explosive Emission Cathode Plasmas in Intense Relativistic Electron Beam Diodes," NRL Memorandum Report 5492 (1985).

16. R. H. Fowler, L. Nordheim, *Proc. Roy. Soc.* **119**, 173 (1928).

17. H. C. Miller, "Values of Fowler-Nordheim Field Emission Functions: $v(y)$, $t(y)$ and $s(y)$," *Franklin Inst. J.* **282**, 382 (1966); and H. C. Miller, "Values of the Electron Emissions Function $v(y)$, $t(y)$ and $\theta(y)$; for $y \geq 1$," *Franklin Inst. J.* **287**, 347 (1969).

18. W. P. Dyke, J. K. Trolan, E. E. Martin, and J. P. Barbour, "The Field Emission Initiated Vacuum Arc. I. Experiments on Arc Initiation," *Phys. Rev.* **91**, 1043 (1953).

19. W. S. Boyle, P. Kisliuk, and L. H. Germer, "Electrical Breakdown in High Vacuum," *J. Appl. Phys.* **26**, 720 (1955).

20. L. B. Snoddy, "Vacuum Spark Discharge," *Phys. Rev.* **37**, 1678 (1931).

21. J. A. Chiles, Jr., "A Photographic Study of the Vacuum Spark Discharge," *J. Appl. Phys.* **8**, 622 (1937).

22. I. Smith, G. Rice, and H. Aslin, "Advanced Flash X-ray System," Air Force Weapons Laboratory Report AFWL-TR-68-113-Vol. I., Kirtland AFB, NM, (1970).

23a. M. Okawa, T. Shioiri, H. Okubo, S. Yanabu, "Area Effect on Dielectric Breakdown of Copper and Stainless Steel Electrodes in Vacuum," *Proceedings of XII International Symposium on Discharges and Electrical Insulation in Vacuum,* Shoresh, Israel, IEEE Cat. No. 86CH2194-9, pp. 65–69 (1986).

23b. S. Y. Ettinger and E. M. Lyman, "Effects of Gas Conditioning on Cathode Surfaces, Field Emission and Electrical Breakdown," *Proceedings of the Third International Symposium on Discharges and Electrical Insulation in Vacuum,* Paris, p. 128–133 (1968).

24. C. M. Cooke, "Residual Pressure and Its Effect on Vacuum Insulation," *Proceedings of the Second International Symposium on Insulation of High Voltages in Vacuum,* Cambridge, MA, p. 181 (1966).

25. R. N. Bloomer and B. M. Cox, "Some Effects of Gases Upon Vacuum Breakdown Initiated by Field Emission of Electrons," *Vacuum* **18**, 379 (1968).

26. R. B. Miller, *An Introduction to the Physics of Intense Charged Particle Beams,* Plenum, New York (1982).

27. G. B. Frazier, "Pulsed Electron Field Emission from Prepared Conductors," *Digest of Technical Papers of the Second IEEE International Pulsed Power Conference,* Lubbock, TX, IEEE Cat. No. 79CHI505-7, (1979).

28. R. D. Evans, C. M. Cooke, and E. R. Berman, "Evolution of Methods for Increasing Vacuum Breakdown Strength," Ion Physics Co. Report C00-4002-080-3, Burlington, MA 01803 (1977).

29. A. S. Gilmoure and R. J. Clark, "Studies on a Laser-Triggered High Voltage, High Vacuum Switch Tube," *Proceedings of the Third International Symposium on Discharges and Electrical Insulation in Vacuum,* Paris, pp. 367–372, (1968).
30. A. S. Gilmore and D. L. Lockwood, "The Interruption of Vacuum Arcs at High D.C. Voltages," *IEEE Trans. on Electron Devices* **ED-22**, 173–180 (1975).
31. W. J. Sarjeant, "High Pressure Surface-Discharge Plasma Switches," *IEEE Trans. on Plasma Science* **PS-8**, 216–226, (1980).
32. J. P. VanDevender, "Vacuum Insulators and Magnetically Insulated Power Flow," Pulsed Power Lecture Series No. 19, Air Force Office of Scientific Research Grant No. 78-3675, published by Texas Tech University, Lubbock, TX (undated).
33. P. N. Dashuk and E. K. Chistov, "Electric Field Configuration for Grazing Discharge," *Sov. Phys. Tech. Phys.* **24**, 687–688 (1979).
34. Yu. L. Stankevich, "Initial Stage of an Electrical Discharge in a Dense Gas," *Sov. Phys. Tech. Phys.* **15**, 1138 (1971).
35. S. T. Pai and J. P. Morton, "A Preliminary Study of the Breakdown Mechanism of Surface Discharge Switches," *IEEE Conference Record of 1982 Fifteenth Power Modulator Symposium,* IEEE Cat. No. 82CH1785-5, pp. 153–159 (1982).
36. I. D. Smith, "Impulse Flashover of Insulators in Poor Vacuum," *Proceedings of First International Symposium on Insulation of High Voltages in Vacuum,* Boston, MA, p. 261 (1964). This work as well as that of other authors dealing with more complex shapes of insulators and electrodes has been reviewed in a more readily available article by R. Hawley, "Solid Insulators in Vacuum: An Invited Paper," *Vacuum,* **18**, 383 (1968).
37. I. Smith, "Tests of a Dielectric-Vacuum Surface Flashover Switch," *IEEE Conference Record of 1982 Fifteenth Power Modulator Symposium,* IEEE Cat. No. 82CH1785-5, p. 160–163 (1982).
38. J. C. Martin, "Fast Pulse Vacuum Flashover," SSWA/JCM/713/157 Atomic Weapons Research Establishment, Aldermaston, U.K. (1971).
39. J. D. Shipman, "Final Report on the NRL Design for an Alternate Diode for PROTO II at Sandia National Laboratory," unpublished (1977).
40. P. Spence, Physics International, Inc., San Leandro, CA (undated) (unpublished).
41. J. E. Boers, "FFEARS-A Digital Computer Program for the Simulation of Laplace's Equition," Sandia National Laboratories Report SC-RR-71-0377 (1971). (See also Ref. 19 in Ch. 3.)
42. S. J. Sackett, "JASON—A Code for Solving General Electrostatic Problems," Lawrence Livermore National Laboratory Report UCID-17814, (1978).
43. C. L. Enloe and R. E. Reinovsky, "Ultra-Violet Induced Insulator Flashover as a Function of Material Properties," *Digest of Technical Papers of the Fourth IEEE Pulsed Power Conference,* IEEE Cat. No. 83CH1908-3, pp. 679–682 (1983).
44. A. A. Avdienko, "Surface Vacuum Breakdown of a Solid Dielectric by Ultrasoft X-rays," *Sov. Phys. Tech. Phys.* **24**, 691 (1979).
45. A. G. Bedrin, I. V. Podmoshenskii, and P. N. Rogovtsev, "Formation of Light-Initiated Surface Discharges," *Sov. Phys. Tech. Phys.* **28**, 403–407 (1983).
46. A. G. Bedrin, I. V. Podmoshenskii, and P. N. Rogovtsev, "High Current Stage of a Light-Initiated Surface Discharge," *Sov. Phys. Tech. Phys.* **28**, 1180–1183 (1983).
47. K. D. Ware, J. W. Mathew, A. H. Williams, P. J. Bottoms, and J. P. Carpenter, "Design and Operation of a Fast High Voltage Vacuum Switch," *Rev. Sci. Inst.* **42**, 512–518 (1971).
48. EG & G Electro-Optics Data Sheet G6003A-1, EG & G, Salem, MA 01970 (1979).
49. General Electric Co. Technical Information Sheet for ZR-7512 Triggered Vacuum Gap, Microwave Tubes Operations, General Electric Co., Schenectady, NY 12305.
50. J. E. Thompson, R. G. Fellers, T. S. Sudarshan, and F. T. Warren, Jr., "Design of a

Triggered Vacuum Gap for Crowbar Operation," *IEEE Conference Record of the Fourteenth Pulse Power Modulator Symposium,* IEEE Cat. No. 80-CH1573-5 ED, pp. 85–91 (1980).

51. F. T. Warren, J. M. Wilson, J. E. Thompson, and T. S. Sudarshan, "Triggered Vacuum Switch Breakdown and Conduction Characteristics," *Digest of Technical Papers of the Third IEEE International Pulsed Power Conference,* IEEE Cat. No. 81CH1662-6, pp. 436–439 (1981).

52. J. Christiansen and C. Schultheiss, "Production of High Current Particle Beams by Low Pressure Discharges," *Z. Physik A,* **290,** 36 (1983).

53. E. Boggasch and H. Riege, "A 400 kA Pulse Generator with Pseudo-Spark Switches," *Digest of Technical Papers of the Fifth IEEE Pulse Power Conference,* IEEE Cat. No. 85C2121-2, pp. 820–823 (1985).

54. G. Mechtersheimer, R. Kohler, and T. Lasser, "Der Pseudofunkenschalter als Ersatz von Thyrotrons und Hochdruckfunkenstrecken," *Verhandl. DPG* (VI) **20,** 985 (1985).

55. D. Bloess, I. Kamber, H. Riegge, G. Bittner, V. Bruckner, J. Christiansen, K. Frank, W. Hartmann, N. Lieser, C. Schultheiss, R. Seebock, and W. Steudtner, "The Triggered Pseudo-Spark Chamber as a Fast Switch and as a High Intensity Beam Source," *Nuclear Instruments and Methods* **205,** 173–184, North Holland, Amsterdam and London, (1983).

56. E. J. Lauer and D. L. Birx, "Tests of a Low Pressure Switch Protected by a Saturating Inductor," *IEEE Conference Record of the Fifteenth Power Modulator Symposium,* IEEE Cat. No. 82CH1785-5, p. 47–50 (1982).

57. R. E. Reinovsky, W. L. Baker, Y. G. Chen, J. Holmes, and E. A. Lopez, "Shiva Star Inductive Pulsed Compression System," *Digest of Technical Papers of the Fourth IEEE Pulsed Power Conference,* IEEE Cat. No. 83CH1908-3, pp. 196–201, (1983).

58. A. V. Grigoriev, P. N. Dashuk, S. N. Markov, V. L. Shutov, and M. D. Yurysheva, "Low-Inductance Megampere-Current Based on Sliding Discharge," *Instrum. and Exp. Tech.* **19,** 1104–1106 (1976).

59. P. N. Dashuk, A. K. Zinchenko, and M. D. Yarysheva, "Erosion of Dielectrics in the Switching of High-Pulsed Currents by a Grazing Discharge," *Sov. Phys. Tech. Phys.* **26,** 196–201 (1981).

60. E. P. Belkov and P. N. Dashuk, "Recovery of Electrical Strength in Gaps After a Sliding Discharge Along the Surface of an Insulator," *Sov. Phys. Tech. Phys.* **25,** 1354–1357 (1980).

5
Solid Dielectric Breakdown Switching

INTRODUCTION

Simplicity, low inductance, high current and voltage operation, as well as some specialized applications (such as, for example, formation of plane wavefront in flat transmission lines for traveling wave excitation of high power nitrogen lasers[1]) make solid dielectric switches useful, practical and sometimes indispensable for high power pulsed applications. These switches are similar to gap switches described in the preceding chapters, except that the solid insulator, usually in the form of thin sheets, is inserted between the electrodes. Figure 5-1 shows an exploded view of such a switch, including a trigger electrode and associated trigger circuit[2]. (An illustration of complete circuitry and switch configuration developed for powering gas lasers is given in the last section of this chapter.) Through exploitation of the solid dielectric (as an insulator with much higher breakdown strength than that of the gases), the switch length is much shorter for a given switching voltage. The operation and characteristics of such switches as well as some applications are described in this chapter.

The high breakdown strength of solid materials contributes not only to low inductance of the switch, because of its short length, but also allows such switches to operate in practice at voltages higher than 1MV. To achieve high operating voltages, the entire switch assembly must be submerged in oil or other insulating liquids (even high resistivity water is suitable when sufficiently fast charging of the electrodes is used). Depending on the type of fluid surrounding the switch, flashover (to which the dielectric surfaces are susceptible) is controlled by both the design of the electrode edges to minimize the enhancement of fields parallel to the surface and by lengthening the flashover surface. These points are elaborated in the succeeding sections. In addition to providing very fast and very low inductance switches, solid gaps have another advantage. In a reasonably laid out system, electromagnetic noise is greatly reduced, since the switch can be buried in the transmission line so that very little of the electric energy is outside

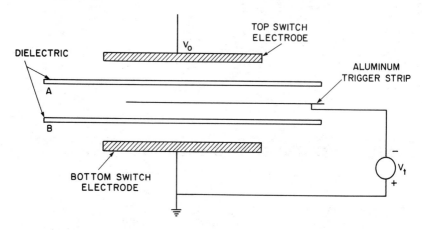

Fig. 5-1. Exploded view of a triggered solid dielectric switch showing main components. Reproduced from Ref. 2.

and it is possible to operate sensitive equipment near systems with 10^{13} A/s current rate of rise[3].

There are two types of solid dielectric switches in addition to a simple self-breakdown switch. One type uses mechanical initiation of the breakdown and electrical triggering is used in the other type. Mechanical initiation can be achieved by using "thumb tack and hammer" operation to puncture partially the dielectric while the electric field is applied across it[4]. Another type of initiation, frequently used, is obtained using detonators or exploding foils to provide a very rapid break of insulation[4, 5, 6]. Komelkov mentions even the use of a high power laser to induce insulator break and initiate a discharge[4].

Solid dielectric switches are used as closing switches or for crowbarring (clamping) a discharge circuit. The latter function is possible, because their resistance is generally very low. As for closing switches, Martin (in Note 4 of Ref. 7) describes an operation using a quite old system which has 40 switches fired simultaneously at 20 kV, producing 4 MA current, rising in 8 ns, i.e., generating current rate of 5×10^{14} A/s. Of course, such switches are good for single-discharge operation. They are, however, of such simple construction that replacement is cheap and can easily be mechanized to achieve a rate of operation of a few times a second[3].

To understand solid dielectric breakdown and surface flashover, reference to material accumulated in the development of high energy capacitors and high voltage and high current pulse transformers provides much of the needed switch design information. One of the more recent surveys of the behavior of solid fibers and sheets has been performed by Sarjeant et al.[8] This work deals with breakdown and flashover dependence on time and on other factors related to solid dielectrics and insulating liquids used in impregnation of capacitors and multi-turn transformers employed in pulse power applications. The survey includes extensive references to detailed data and analysis and provides useful design scalings discussed in other parts of this chapter.

BREAKDOWN STRENGTH OF SOLIDS

In contrast to the breakdown strength of gases under pulsed conditions (discussed in Chapter 3), which may reach 100 kV/cm at suitably high pressures, solid (and liquid) dielectrics are characterized by much higher breakdown strength, typically, in the 1 MV/cm range. Following Ref. 8, three characteristics of solids (which are also common to liquids and are further discussed in Chapter 6) used in pulsed applications are:

1. Time-dependence of breakdown strength in the approximate range of 10^{-6} to 10^{-9} s.
2. Weak decrease in breakdown strength with increasing area of the applied electric field in the range of 1 to 10^4 cm^2.
3. Weakly dependent breakdown strength which increases as the insulator thickness decreases from about 1 cm to the thinnest sheets available commercially, about 10^{-4} cm.

Sarjeant et al.[8] describe qualitatively the solid dielectric breakdown in terms of statement 1 above. At a very early time, intrinsic breakdown electric field is high. As the time of an applied field extends, streamers leading to lowered breakdown begin to develop. At an even later time, thermal erosion and electromechanical tracking take place, usually in association with repetitive charging of the dielectric. Smith and Martin[9] have tested large volumes of several plastics (with pulse charging time of about 10^{-6} s) to obtain quantitative scaling dependence, integrating the area and thickness effects of statements 2 and 3, above, into dependence of breakdown on volume. Figure 5–2 is reproduced from Ref. 9, where it is noted that the results are lower than those quoted in the literature for "intrinsic" breakdown. Similar data for Mylar on breakdown strength versus thickness, quoted in Ref. 8 after du Pont Manufacturing data on electrical properties of Mylar polyester films (February 1963) shows that very thin sheets of Mylar (5×10^{-4} cm) have breakdown strength of 8 MV/cm. Assuming those measurements to be for areas of about 1 cm^2, i.e., assuming a test volume of about 10^{-3} cm^3, the data in Ref. 9 scales to the values given in

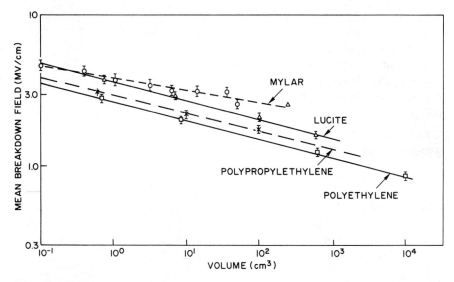

Fig. 5-2. Dependence of mean breakdown field of four dielectric sheets on dielectric volume subjected to electric field stress. Data shown in the figure are from Ref. 9.

Ref. 8. For completeness, it should also be mentioned that as a solid dielectric switch is built, stacking of sheets to a given thickness leads to a somewhat higher breakdown level than for the same thickness made of a single layer of plastic[10, 11].

The streamers formed in the breakdown of solids have short transit times for establishing a discharge channel. The breakdown value is nearly independent of pulse duration (down to a few nanoseconds). A practical expression for the breakdown is given in Ref. 12:

$$EV^{1/10} = K \qquad (5\text{-}1)$$

where the electric field, E, is in MV/cm and volume of the stressed dielectric, V, is in cm^3. The value for the constant K is given as 2.5 for polyethelene and Teflon, 2.9 for polypropylethelene, 3.3 for Lucite, and 3.6 for thick Mylar.

Furthermore, for solids there is a reduction in the breakdown field for repeated pulses[12] (with number of charging cycles before failure called "life"):

$$\text{Life} \propto (E_b/E_{op})^8 \qquad (5\text{-}2)$$

that is, it depends on the strong power of the ratio of the breakdown field, E_b, to the operating field level, E_{op}. Finally, both polyethelene and polypropylethelene show a polarity effect when pulse-charged in divergent geometries[12]. For example, when one of the electrodes is a sphere and it is charged negatively with respect to a flat electrode, breakdown occurs at twice the level as compared to that associated with the opposite charging. This is further illustrated later in this chapter.

In addition to direct breakdown of the solid dielectric, it is possible to initiate the discharge between the electrodes also with surface discharges. The dielectric surfaces outside the electrodes, seen in Fig. 5-1, provide a path for discharge from one electrode to the other. In general, immersing a solid dielectric and electrodes of such a switch in an insulating liquid helps to prevent the occurrence of such discharges. Transformer oil is a good liquid for this purpose; in pulsed systems, water is also used for this purpose. Because the dielectric constant of water is very high ($\epsilon = 80$), it is more difficult to prevent surface breakdown when water is used. Acceptable methods have been developed for pulse voltages of about 4 MV in laboratory systems[12]. Considerable discussion of methods for preventing surface discharges at a level of up to a few hundred kilovolts is discussed by Komelkov[4], including the cases in which the applied field is an oscillating type.

In submerging the dielectric sheets in a liquid to reduce flashover, it is easy to entrap air layers and bubbles between the solid sheets. These, because of lower dielectric strength, will breakdown at a field value different from the solid. The fields in each component of the "sandwich" dielectric are[8]:

$$E_a = V_0/[d_a + d_s (\epsilon_a/\epsilon_s)] \qquad (5\text{-}3a)$$

$$E_s = V_0/[d_s + d_a (\epsilon_a/\epsilon_s)] \qquad (5\text{-}3b)$$

where subscripts a and s refer to air and solid breakdown fields (E) respectively; and d, thickness, with ϵ the dielectric constant.

MECHANICALLY INITIATED SWITCHES

Mechanically operated solid gaps have been used for a long time and for many dc applications provide probably the fastest rise time[12] when used in low impedence circuits. Komelkov indicates[4] that such gaps were first proposed by him in 1956[13], involving the use of an exploding wire to generate a strong shock for cracking the dielectric between the main electrodes, thus initiating the breakdown and subsequent discharge of the main circuit. Various authors have extended the use of shocks employing precisely triggered detonators for parallel switching and crowbar operation.

Pin-Triggering

The principle of pin-triggered electrodes is very simple. A relatively sharp metal needle or rod is driven against the insulating sheet with enough force to penetrate it. As the penetration occurs, the applied electric field increases across the remaining insulation to a point where the breakdown occurs. A very small volume of the dielectric associated with the high electric field in the penatration area suggests on the basis of the data, such as that given in Fig. 5–2, that breakdown occurs at about 8 MV/cm. The mechanical force also deforms the electrodes to form a good feed to a very short plasma (discharge) channel[12], explaining its very low resistance.

The force driving the breakdown-initiating pin can be a simple hammer (appropriately insulated form the operator), a pneumatically operated mechanism, or an electromagnetic device. Roger and Whittle[14] have developed the last device for switching currents up to 600 kA. Its operation and performance are summarized by Komelkov[4]. The switch consists of an aluminum hammer of 3.8 cm diameter driven by a pulsed coil. The hammer is tipped by a steel cup matching the copper rivet (hollow cylinder of 1.27 cm diameter). The rivet is lodged in one of the flat electrodes. Between the electrodes, a polyethelene sheet of 0.5 mm thickness provides the insulation before switching. The electrode on the opposite side of the hammer also has a hole backed up by a steel cone designed to fit inside the rivet, as the rivet is driven through the polyethelene. The stroke time is 40 μs for 0.5 cm travel distance. When the rivet moves rapidly, the resistance 1 μs after closure is 1.7 microhms. At 300 μs, when the current reaches 600 kA, the voltage drop across the switch is a few volts.

Pressure-Triggering Using Exploding Conductors

Very similar switching characteristics to those using mechanical puncture of the insulation is obtained by use of localized high pressure impulses. Each impulse is generated by the rapid resistive heating of wire, or foil, in a confined volume, to above-vaporization temperature. James, et al.[16, 17] have developed a 1 MA solid gap switch, holding off 100 kV with a polyethelene sheet 0.5 cm thick. Two varieties of the switch are shown in Fig. 5–3. The polyethelene insulation is punctured by a trigger discharge through a 0.03 cm hole in an auxiliary insulation (thin polyethelene sheet placed between copper foil trigger electrodes). With a risetime of 1 μs from a 20 kV, 1 μF capacitor storing 200 J the trigger pulse initially breaks down through the 0.03 cm hole capillary leading to a pressure build-up that punctures the main dielectric. The main discharge is switched in 1.0 ± 0.1 μs. The trigger mechanism allows the switch to operate virtually independently

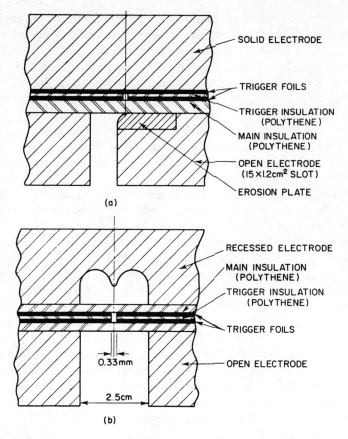

Fig. 5-3. Symmetric (bottom) and assymetric (top) configurations of solid dielectric switches using high pressure, generated by exploding conductors, to disrupt mechanically the integrity of gap insulation. Reproduced from Ref. 17, courtesy of U.K. Energy Authority, Culham Laboratory.

of the switch voltage so that these switches can operate also as crowbar switches.

When operating at peak currents of several megamperes and switching 1000 coulombs, the electrode assembly and support structure must be sufficiently robust to withstand high magnetic forces and contain the thermal energy dissipated in the switch arc resistance. Typically, magnetic forces of several hundred tons can be encountered in megajoule discharges; 10–20% of this energy may go into the switch area. To avoid the pressure build-up and to remove polyethelene vapor or gas and electrode erosion products, a slot or hole is cut in one of the electrodes as shown in Fig. 5-3. It is advantageous to place the breakdown point (determined by the position of the capillary in the trigger insulation) opposite the center of the slot and to add a recess of the same shape in the opposite electrode, as shown at the bottom of Fig. 5–3. This configuration prevents the erosion and contamination of the insulated electrode surface at current levels up to 2 MA, so that the switch can be used for subsequent discharges. In crowbar applications, requiring low arc voltage, the switch configuration at the top of Fig. 5–3 is preferable. The arc length, and therefore its resistance, is minimized, but the electrode erosion becomes severe. To alleviate the effects of the erosion, James[17] recommends the use of removable plates fitted to the electrode near the breakdown point.

One difficulty in working with the low resistance spark gap is the fact that a difference in time of breakdown of each half of the insulation electrically stresses the unbroken part of insulation to full applied switch voltage that could break down in other parts of the switch surface. To prevent that from occurring, additional sheets of insulation with a large hole, equal in size to the slot, should be added to the switch sandwich[16]. The voltage drop across the switch shown in Fig. 5–3 is only 250 V at 10^6 A.

Another method of providing sufficient explosive force that can cause electrical breakdown in a solid dielectric switch has been suggested by Barbini[6]. A high power laser for initiating switching by focusing the beam within or on the transparent or non-transparent dielectric membranes, although quite expensive to use, provides the means of totally (electrically) isolating the trigger from the main discharge circuit. A laser can be used to explode relatively thin membranes in 10^{-7} seconds, so that parallel switching of several gaps can be contemplated.

Metal-to-Metal Switch Closing

In those applications where very low switch resistance, of the order of 10 microhms, is needed or where the switch must close when no voltage is being applied across the switch, metal-to-metal contact can be achieved by a method similar to that described in the preceeding section. Figure 5–4 illustrates such a switch as described in Ref. 17. In this case, an aluminum metal plate connected to one electrode is rapidly displaced so that it cuts the solid insulation and becomes embedded in a groove of the initially insulated opposite electrode. Metal-to-metal contact of about 10 cm in length is established in the switch. Exploding foil is again used to provide the explosive pressure to deform the aluminum plate. As James[17] indicates, this is a preferred method to an electromagnetic hammer, because the large weight of the hammer leads to a rather large jitter (about 1 microsecond) whereas the exploding foil trigger[19, 20] reduces the closing time to 3.0 μs

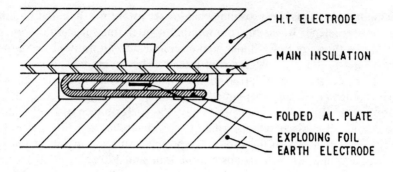

H.T. ELECTRODE

MAIN INSULATION

FOLDED AL. PLATE

EXPLODING FOIL
EARTH ELECTRODE

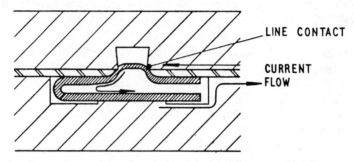

LINE CONTACT

CURRENT
FLOW

Fig. 5–4. Solid dielectric metal-to-metal contact switch shown before (top) and after (bottom) operation. Reproduced from Ref. 17, courtesy of U.K. Energy Authority, Culham Laboratory.

with a jitter time of 0.3 μs. Typically, foil current for such performance is 80 kA rising in 1 μs from a 3 kJ trigger capacitor bank[17].

When metal-to-metal switching is used to switch voltages which can sustain an arc before metal contacts are established, substantial erosion of the electrodes can occur to prevent the switch from multiple use. What is often employed to overcome this deficiency is another parallel switch (which could be another solid switch or a gas gap) providing the initial switching, dropping the voltage across the metal-to-metal switch, so that when contact occurs, only very low voltage appears across the contacts. Dokopoulos et al.[18] have developed such switching configuration, employing the field distortion gas spark gap with ±20 ns firing jitter. A metal-to-metal switch takes over the current-carrying function at about 2 μs, with a resistance of 10 microhms (contrasting with the spark gap resistance of a few milliohms). This switch has been used in the range of 20 to 60 kV for switching of currents up to 450 kA. To avoid prefiring at 60 kV, two mylar sheets, each 0.35 mm thick, are used. This leads to somewhat longer jitter time than that obtained with polyethelene. Regarding the minimum jitter time, it should be noted that the authors of Ref. 18 also make reference to an unpublished work[19] which reports that 0.1 microsecond jitter has been obtained.

A more complete description of the metal-to-metal switch design, with performance essentially the same in both start (closing) and crowbar operation, is given in Ref. 20. Specifically, the performance of the initiating exploding foil is discussed in terms of reproducibility and jitter time. Operation of 500 kA with 0.4 mm polyethelene insulation in the switch gap has resulted in 2.4 microseconds closing time and 170 ns jitter. To achieve such performance, Al or Cu foils 15 micrometers thick are used. The foil must be shaped to include short (6 mm) narrow section intended to explode first. This way, switch closing occurs in the same place relative to the main electrodes. The foil is exploded by current from a 0.9 kJ (7 microfarad, 16 kV) capacitor with trigger circuit inductance of 350 nH. Under these conditions, the foil explodes in 250 ns. For this igniter bank, the switch delay time (i.e., time to breakdown) remains constant for voltages between 20 and 15 kV. As the voltage drops further, the switch time becomes longer. Using higher trigger energy has resulted in increased electrode erosion, without an improvement in performance. The switch delay time increases approximately linearly with insulation thickness. As the thickness increases from 0.4 mm to 1.8 mm, switching time changes from 2 μs to about 8 μs, with jitter not exceeding ±200 ns. Air pockets in the switching region lead to a severe increase of switching time (and decrease of the contact area). Introducing a 0.4 mm air gap in addition to the 0.4 mm polyethelene sheet between the electrodes doubles the switching time. In order to obtain reproducible results, the switching parts have ben pressed together with a constant force of 5 tons. Finally, the jitter could be reduced to ±70 ns by suppression of any pockets that could vent the high gas pressure resulting from the exploding foil.

Detonator-Triggered Start and Crowbar Switches

Major development of the solid gap switches using detonators was performed at the Naval Research Laboratory in the early 1970's[5], following the work of Dike et al.[21] Chemical detonators have definite advantages for some capacitive storage and inductive storage applications. They are particularly interesting for some high voltage switching applications, where it may be necessary to hold-off initially 200 to 300 kV and then to switch accurately at very low voltage (15 kV) as discussed in conjunction with methods of forming meter-long guided discharges in the atmosphere[22]. A circuit used in such experiments is described under "Applications" near the end of this chapter. Systems requiring large numbers of parallel-connected starting and crowbar switches are served conveniently by detonator-initiated switches, with their good firing simultaneity and reliability[5]. Because cost per shot is high (several dollars per detonator), the detonators are used in systems with a small number of discharges per day. However, their simplicity of operation, and recently employed optical trigger signal isolation[23], and availability of detonators in a wide range of sizes suggests wide use for such switching.

The high concentration of energy associated with the detonator-initiated switches, similar to that generated by exploding foil, suggests that the designs of detonators and foil-triggered switches can be essentially the same. The detonator-initiated switches of Fig. 5–5 is based on the principles of the foil-triggered switch illustrated in Fig. 5–3. It is a switch for closing via an arc formed through a hole punched in the insulating sheets by the explosive pressure of the detonator directly in contact with the sheets. The high pressure gas from the detonator and intense fireball, created by the high voltage arc, is released into an in-line vent with a diameter approximately the same as that of the detonator. Removable electrodes provide a safe configuration for a simple and safe reloading of the detonators in a separate detonator handling facility[5].

Ten switches of the type shown in Fig. 5–5 are discharged in parallel to operate a 3 MA capacitor bank, storing about 0.5 MJ, into an inductive load. Figure 5–6 is a schematic of the main discharge circuit as well as the triggering circuit for the detonators. The detonators are located on the high

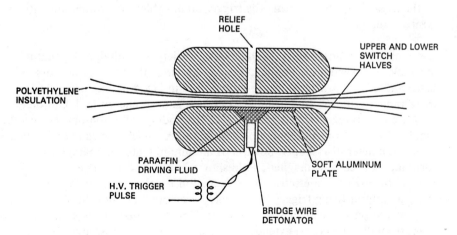

Fig. 5–5. Detonation-initiated starting switch design as modified by M. Raleigh of the Naval Research Laboratory from Ref. 5.

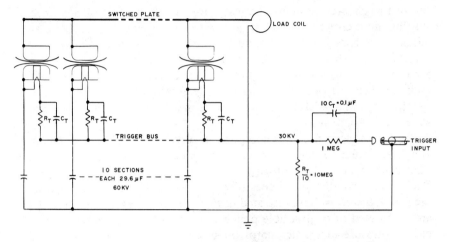

Fig. 5-6. Detonator-triggered circuit for 60 kV capacitor bank using 10 parallel switches to provide 3 MA operation. Reproduced from Ref. 5.

voltage (60 kV) plates of each capacitor section, allowing easy preparation of switches for firing. This required a development of the trigger circuit also shown in the figure. The divider network reduces the trigger bus to one-half of the capacitor voltage, and when the system is charged draws about 5 mA current. On command, the trigger bus is pulsed to ground, producing a high firing current to each detonator. The detonators perform well with a voltage range from 10 to 60 kV, requiring only an adjustment of the trigger gap (terminating the trigger input cable) to accommodate the desired voltage.

Detonator Characteristics. There is a variety of suitable detonators available for switching applications. Triggering of these detonators is listed and their characteristics are summarized in Fig. 5-7. Explosive charge mass ranges from 60 to 320 mg, releasing chemical energy in 1 kilojoule range, comparable to that released by the exploding foil initiators. For example, if metal-to-metal contact is desired, a 200 mg charge can accelerate a 2 cm diameter disc-shaped portion of aluminum plate (in analogy to distortion produced by a magnetic field shown in Fig. 5-4) to velocity of 0.25 mm/μs through the dielectric. The "function time" shown in Fig. 5-7 is the time after a firing trigger has initiated a detonation until the shock arrives at the output end of the detonator. This time may be as small as 2 microseconds, or may be extended to several tens of microseconds. Firing jitter may be as low as ±25 ns. The cost of the detonator remains roughly unchanged from that quoted in Fig. 5-7 (prepared in 1973).

	BLASTING CAPS	CONDUCTIVE MIX	SHAPED CHARGE	MED. PRECISION	HIGH PRECISION
TYPE OF REACTION	DEFLAGRATION DETONATION	DETONATION	DETONATION, SHAPED CHARGE	DETONATION	DETONATION
INITIATOR	HOT WIRE	CONDUCTIVE MIX	*EBW	*EBW	*EBW
CROSS SECTION	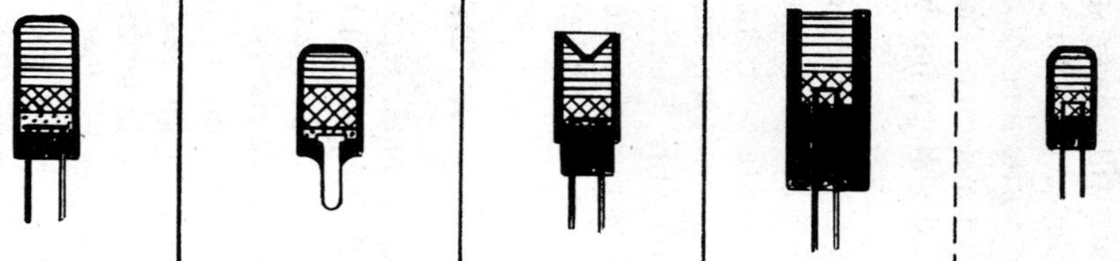				
▥ BASE CHARGE	320 mg PETN	19 mg HMX	60 mg PETN	124 mg RDX	40 mg PETN
▧ PRIMER CHARGE	PETN	HMX	PETN	93 mg PETN	24 mg Tetryl
▦ IGNITION CHARGE	Lead Azide	28 mg Lead Azide	NONE	NONE	NONE
MIN. TRIGGER ENERGY	48,000 ergs	10 ergs	$\frac{di}{dt} > 1$ kA/μs	$\frac{di}{dt} > 1$ kA/μs	$\frac{di}{dt} > 1$ kA/μs
TYPICAL TRIGGER	5 A	2 V, > 10 mA	800 A	800 A	800 A
TYPICAL FUNCTION TIME	> 5 μs	4 μs	4 μs	3 μs	2 μs
FIRING DEVIATION	> .1 μs	.125 μs	.125 μs	.125 μs	.025 μs
APPR. COST	< $1.	$4. - $7.	$15.	$3. - $5.	$15.

*EBW = EXPLODING BRIDGE WIRE

Fig. 5–7. Characteristics of detonators available for use in triggering of solid dielectric switches. Reproduced from Ref. 5.

The Reynolds Industries, Inc.*, RP 80 detonator is used extensively in conjunction with the NRL 0.5 MJ capicitor bank and in the inductive storage applications. This relatively powerful detonator was selected because of its short (±200 ns) firing deviation time, cost, and availability. It was installed in a 4340 steel breech-block, heat treated to a tensile strength of 250 kpsi. This was needed because materials with lower strength can be deformed by the explosive forces, causing unacceptable trigger jitter after only a few shots. Dependence of the characteristic time of contact formation of the metal-to-metal switch and of arc formation of the closing (start) switch on insulation thickness and on applied voltage has been studied in some detail and is reported in Ref. 5.

As noted in Fig. 5-7, normal detonation triggering energies vary from very sensitive (10 joules) for "conductive mix" devices to highly insensitive EBW (Exploding Bridge Wire) devices, requiring fast current rise to explode a bridge wire needed to set off the explosive. However, since under some conditions even the least sensitive detonator may fire, Ford and Young stress that it should be assumed that each detonator may explode and safe handling procedures, such as those in Ref. 5, must be instituted.

SELF-BREAK AND ELECTRICALLY TRIGGERED SWITCHES

In the early 1960's, in an attempt to provide a simple reproducible switch for work in the 100 kV region with a current change rate of 10^{14} A/s, Martin[3] and collaborators developed highly practical solid dielectric switches. The configuration of such switches, using polyethylene sheets, is illustrated in Fig. 5-1. To localize the breakdown area, the sheets are stabbed with sewing needles, making a sharp depression part way through the dielectric. When this procedure was developed, it was reasoned that the air in the void would break down early in the charging phase at voltage gradients of two orders of magnitude less than in the polyethylene, so that at the time of closure the stab would be equivalent to a sharp metal electrode. (When such switches are submerged in water or oil, air remains entrapped in the depression.) By controlling the stab length, it is possible to vary the switch operating voltage easily. Figure 5-8 shows the dependence of the breakdown voltage for a single sheet between the electrodes on the normalized thickness of the dielectric, b/a, as defined in the figure.

The breakdown voltage curves (for 1.5 mm polyethylene sheet) in Fig. 5-8 are seen to depend weakly on the charging time. The lower curves correspond to the negative voltage applied to the stabbed side. When the volt-

---

*San Ramon Facility, 3420 Fostoria Way, San Ramon CA, 94583 (Tel.: 415-837-0567).

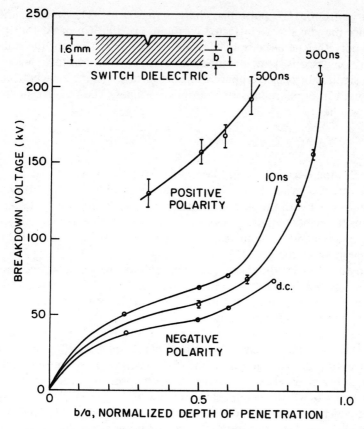

Fig. 5-8. Breakdown voltage of stabbed 1.6 mm thick polyethelene sheets. Reproduced from Ref. 3.

age is reversed, the breakdown level is higher by a factor of 2. When using a single stab, the standard deviation in the breakdown level was found by Martin[2, 7] to be about 10%, and as the number of stabs was increased (with a mechanized stabber) to 50, the standard deviation was reduced to 2-3%. For multiple stabs, the operating range of the switch was reduced.

A different version of the controlled breakdown switch was also tested[2, 7], up to 500 kV when the stab was replaced by a buried small bearing in 0.62 mm thick polyethylene. This relatively crude switch achieved a standard deviation of 8% in breakdown values. In those instances in which the switch inductance could be measured or deduced, the inductance of stabbed and ball bearing switches was within about a factor of 2 relative to the calculated value, typically 2-5 nH for a single breakdown arc driven by a discharge from a 1 Ω pulse line charged to 200 kV. The value of the inductance may be slightly affected by the use of thin Cu or Al foils taped to the polyethelene because of the vaporization in the arcs of the stab.

Work on gaseous spark channels, discussed by J. C. Martin[12], has shown that a significant resistive phase can exist in fast risetime switches such as those used in discharging strip transmission lines. The resistive phase for solid switches is[12]

$$\tau_R = 5 \, Z^{-1/3} \, E^{-4/3} \text{ nanoseconds} \qquad (5\text{-}4)$$

where E is the electric field in MV/cm and Z is the impedance of the pulse line driving the switch.* For the case mentioned above, $Z = 1$ Ω and the field associated with the stabbed switch is perhaps as high as 5 MV/cm, τ_R is about 0.5 nanoseconds. Thus, the inductive rise time component, L/Z, of 2–5 nanoseconds, dominates the risetime of the switch for the 1 Ω pulse line operating at 200 kV.

To reduce further the risetime associated with the solid switches, multiple-channel switching can be utilized. Multiple-channelling in the switch is defined, following Martin in Ref. 7, as channelling in which multiple discharges occur between the continuous electrodes rather than in separate switches fired in parallel by means of transit time isolation. In the multiple-channel switch, while such isolation may play a small part, the main effect is that before the voltage across the first channel can be substantially reduced, other channels start closing. In such cases, it is the inductive and resistive phases which are important in allowing very brief intervals to occur during which multichannel operation can occur. For example, in a 200 kV switch with standard deviation in breakdown voltage of 2.5%, i.e., 5 kV, the combined inductive and resistive phase of 5 ns (mentioned in the example above) would require the charging rate be 10^{12} V/s for multichannelling to take place. Thus, 200 kV must be applied to the switch in 0.2 μs. To achieve such rates in capacitive storage systems with significant energy, energy is transfered from the initial storage capacitor to a capacitor (usually a transmission line of low impedance) capable of rapid discharge which is made possible by use of the solid dielectric switch.

If it is necessary to work with slower charging times, low switch inductance, resulting from multichanneling, can be achieved by triggering of the solid dielectric switch. (Once triggering techniques are available, the triggered switches then can be applied for other purposes, as in voltage multiplying systems where a series of charged transmission lines, each equipped

---

*In Chapter 6, a different convention is followed, using mks units for τ_R and E.

with a precision triggered switch, is connected in series to add all the output voltages.) Figure 5-1 shows how a triggered switch can be put together and operated. Two sheets of polyethelene with multiple stabs, with a metal foil trigger electrode between them, are sandwiched between the main electrodes. The trigger electrode is connected to a pulser supplying the potential V_t across one of the main switch sheets floating at a potential somewhere between ground and V_0. As suggested in Ref. 3, the triggered voltage can arise as a result of capacitive division of the potential applied across the main electrodes—and therefore it is very easy to construct such switches. With the use of a knife edge, or stabs in the polyethylene (for better control) selected to cause a single-channel breakdown, the potential across the top half of the switch increases toward V_0. If number of stabs of appropriate length (so that no breakdown occurs at, say, $V_0/2$) are included in the top sheet, the extremely fast rise from $V_0/2$ to V_0 will result in formation of multichannels in the top half of the switch. This in turn will apply a full potential across the bottom half of the switch, causing it to multichannel. The latter action requires that the initial breakdown at the trigger site be isolated by transit time or inductance from the main discharge path.

In the practical application of such switches, considerable balancing is needed. This is achieved by observations of breakdown voltages and times of each of the gaps separately without introducing other gaps (i.e., stabs) into the switch. Further appreciation of the type of balance that is needed for good switching can be gleaned from the consideration of a specific example discussed in the next section.

The ideas and their practical embodiment have been patented by Martin[24] and by Martin and Smith[25]. Several unpublished notes and various laboratory tests have dealt with the application of such switches for use with d.c. charged capacitor banks to generate 10 MA discharges and in pulse charged systems up to 500 kV.[7]

APPLICATIONS OF SOLID DIELECTRIC SWITCHES

Two rather elegant applications of the solid dielectric switches have been developed by Shipman[26] and Greig and their collaborators.[22] Those applications are selected to show the versatility of this type of switch rather than to show typical applications. In the first application, a method of launching an electromagnetic wave in a flat-plate transmission line was developed to provide a 500 kA excitation current in nitrogen, neon, or hydrogen gas lasers that effectively travels from one end of the laser to the other with a velocity matching that of a stimulated emission. The description of this switch is instructive for designing switches for a variety of other applications where large currents with exceedingly fast risetimes are needed. The second application, developed in conjunction with studies of guided discharges (that is, the induction of current discharge in the atmosphere along a predesignated path)[22] provides a method for connecting high voltage and low voltage power sources to a common load with the controlled time separation between discharges from each source being as short as 1 μs. This method allows the development of complex current waveforms in loads designed, for example, to simulate natural lightning discharges.

The device for generating the traveling wave excitation is shown schematically in Fig. 5–9. It is a flat polyethelene dielectric Blumlein generator with a slot down the middle of the top plate which contains the lasing gas. The two halves of the top plate are charged simultaneously in 1 ns to -80 kV relative to the bottom (ground) plate. The seven switches between the top and bottom plates at the left edge of the generator are designed to close in sequence with a closure velocity, down the line of switches as shown in Fig. 5–9, equal to that of light in a gas medium. This results in the formation of an electromagnetic wavefront which is nearly straight by the time it reaches the laser cavity. The front crosses the gas column at an angle such that the intersection point moves down the column with a phase velocity equal to the velocity of light in the gas. The lengths of the seven transmission line cables between the trigger point and each switch foil must be adjusted to trigger each switch at a proper time. For the precision timing needed here, commercial coaxial transmission cable (with polyethelene dielectric) cannot be cut to length by assuming a uniform transmission velocity. The differences in transmission time between each cable are measured and adjusted using a time domain reflectometer with a 50 picosecond resolution.

Each switch consists of a solid dielectric sandwiched between the generator flat plates. It uses a trigger electrode of aluminum foil 0.025 mm thick and 0.6 cm wide between a Mylar sheet of 0.25 mm thickness nearest the top negative plate and a Mylar sheet of 0.125 mm thickness nearest the bottom positive or grounded plate. At each of the seven switch positions, 5 cm × 5 cm copper-tungsten alloy electrodes are inserted in the top and bottom flat aluminum plates in order to reduce erosion due to high currents. Each switch area is compressed with a 6.8 kg weight.

The schematic circuit for the complete trigger system is shown in Fig. 5–10. (Although only one Mylar switch at the edge of the flat plate line and its trigger cable are shown, there are, of course, seven as shown in Fig. 5–9.) The top plate of the flat line in Figure 5–10 is resonance charged to $-V$ volts from a commercial capacitor through a low inductance in about 1 μs. While the top plate is charging, the potential of the trigger foils is kept at one-third of the potential of the top plate by capacitive division so that no intense fields will exist at the edges of each foil, because at $V/3$

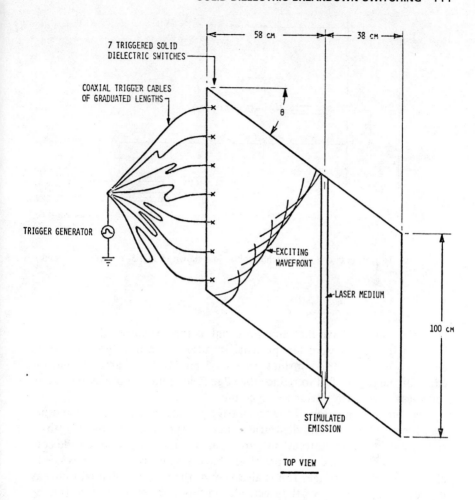

TOP VIEW

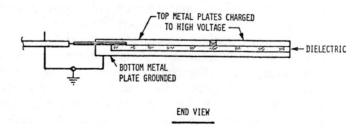

END VIEW

Fig. 5-9. Low impedance pulse line for applying a traveling wave voltage pulse to a gas laser. The pulse line is controlled by an array of switches firing sequentially from top to bottom. Figure provided by J. D. Shipman of the Naval Research Laboratory.

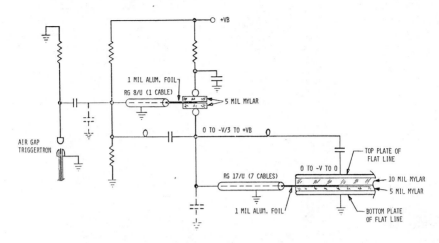

Fig. 5-10. Triggering circuit for switching pulse line shown in Fig. 5-9. Figure provided by J. D. Shipman.

volts the foil will have the same potential as that which would exist at its location if the foil were not present. In addition, a little unpolymerized epoxy with a dielectric constant between 5 and 10 is smeared around the edges of the trigger foil to reduce the edge fields (due to not having a zero-thickness trigger foil) prior to triggering.

The common junction of the seven trigger cables is connected to another very low inductance solid dielectric switch, as shown in Fig. 5-10, with its trigger cable connected to an air gap trigatron and its upper electrode connected to a low inductance, positively biased, capacitor. In this solid dielectric switch, the trigger foil is also kept at the proper potential (midway between the potentials of the electrodes in this case) prior to triggering by resistive and capacitive division. This switch has the electrodes and switch sandwich submerged in transformer oil which serves as an insulator.

The triggering sequence is as follows: When the top plate reaches its peak value of V volts, a trigger generator is used to fire the air gap trigatron which has $V/6$ volts across it resulting from the resonance charge and the rather intricate capacitive division shown. The signal from this closure through the RG8/U cable triggers the single low inductance Mylar switch (using a sheet 1.2×10^{-2} mm thick) submerged in oil. This Mylar switch provides the common fast rising trigger signal (30 kV with a 5 ns risetime) sent into the seven coaxial cables of increasing length to close the seven final Mylar switches. Each Mylar switch is triggered by the signal from the 50 Ω coaxial cable driving the potential of the foil positive to a value near the potential of the bottom plate. This sets up an intense field at the edges of the trigger foil starting ionized streamers which puncture through the 0.25 mm Mylar and rapidly make a low resistance connection with the top plate. The impedance of the connection is very low compared to the 50 Ω trigger cable so that the trigger foil potential is changed rapidly to that of the top plate. This sets up an intense field at the edge of the foil again, starting streamers which puncture through the 0.125 mm Mylar sheet to the bottom plate, completing the closure of the switch between the top and bottom plates. This process results in several punctures per switch at the edges of the trigger foil. The absolute firing time jitter of these switches has not been measured directly, but in order for the traveling wave excitation to work as well as it does, the jitter would have to be small compared to the transit time between switches, which is about 0.50 ns.

The switches are submerged in high resistivity water in the tank containing the pulse line in order to prevent surface discharges. Each of the seven switches can be considered to switch a section of the flat plate line with an impedance of about 3 ohms. The inductance of each switch is only 0.3 nH so that the effect of inductance on risetime is negligible. When Eq. 5–4 is applied to the case with one conducting channel per switch, $\tau_R = 1.4$ ns and the 10% to 90% risetime is 3 ns. The experimentally observed current risetime at the center of the gas column is measured as 2.5 ns. This risetime is lower than

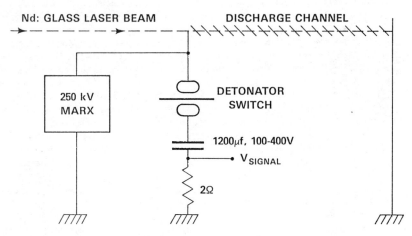

Fig. 5–11. Illustration of a detonator-triggered switch used for holding off 250 kV potential during the discharge of Marx capacitor bank through the channel. The switch is fired to add a high current post-discharge through the channel in series with 2 Ω load. A 1200 μF capacitor serves as a source of current. Figure provided by M. Raleigh of the Naval Research Laboratory.

that calculated because of the formation of multiple channels in each switch. This device as built with solid dielectric switches is, of course, a single-shot system since the switches cannot be reused.

Solid switches with detonator triggering, much simpler compared to those just discussed, have been employed by Murphy et al.[22] as an isolating switch. The circuit shown in Fig. 5-11 is designed to impress a high voltage pulse to a laser-predesignated channel to be followed by a low voltage, long duration discharge. The isolation switch prevents the full Marx generator potential from appearing across the low voltage source. After Marx energy is dissipated, the detonator switch connects the 1200 μF capacitor to the discharge channel. Firing of the detonator blisters an aluminum plate, which then pushes the polyethelene insulation into the relief hole and forms a very low resistance metal-to-metal contact. The switch employs six 0.25 mm thick polyethelene sheets and a Reynolds Industries, Inc., RP-80 detonator. This switch closes in 20 ns, after the detonator pulse, with a jitter of about 1 μs. The sheets have 1000 cm^2 area and are surrounded by SF$_6$ atmosphere to prevent flashover at 250 kV applied for 1 μs.

REFERENCES

1. J. D. Shipman, "Traveling Wave Excitation of High Power Gas Lasers," *Appl. Phys. Lett.* **10**, 3 (1967).
2. T. R. Burkes, *A Critical Analysis and Assessment of High Power Switches,* Naval Surface Weapons Center, Dahlgren, VA, Report NP 30/78 (1978) (same as Ref. 3, Ch. 3).
3. J. C. Martin, "Notes on Some Solid Dielectric Switches," SSWA, AERE, Aldermaston, U.K. (1962) (unpublished).
4. V. S. Komelkov, Ch. 6, "Technology of Large Impulse Currents and Magnetic Fields", translated by Foreign Technology Division, Report No. FTD-MT-24-992-71, from *Tekhnika Bolshikh Impulsnykh Tokov i Magnitnykh Poley,* Atomizdat, Moscow (1970) (same as Ref. 3, Ch. 4).
5. R. D. Ford and M. P. Young, "Chemical-Detonator Solid-Dielectric Switches for Starting and Clamp Applications," *Proceedings of the Eighth Symposium on Fusion Technology,* Commission of the European Communities, Luxemburg pp. 377–390 (1974).
6. S. Barbini, Fourth Symposium on Engineering Problems in Thermonuclear Research, Frascati, Italy, 1966 (unpublished). The record of the Fourth Symposium consists only of the abstracts, published by Laboratorio Gas Ionizzati, EURATOM-CNEN, Frascati, Italy (1966).
7. C. Baum, ed., "Pulsed Electric Power Circuit and Electromagnetic System Design Notes, PEP4-1", Air Force Weapons Laboratory (Albequerque, NM) Report AFWL-TR-73-166, (1973) (same as Ref. 5, Ch. 3).
8. W. J. Sarjeant, G. J. Rowhein, G. H. Mauldin, K. J. Bikford, T. E. Springer, and T. R. Burkes, "Polymer Laminate Structures," *Appl. Phys. Comm.* **3** (1 and 2), 83–167 (1983).
9. I. D. Smith and J. C. Martin, "D.C. and Pulse Breakdown of Thin Plastic Films," Note SSWA/IDS/6610/105, AWRE, Aldermaston, U.K., (1965) (unpublished).
10. J. C. Martin, "Pulse Life of Mylar", Note SSWA/JCM/6611/106, AWRE, Aldermaston, U.K. (1966) (unpublished).

11. J. C. Martin, "Pulse Breakdown of Large Volume of Mylar in Thin Sheets", Note SSWA/ JCM/673/27, AWRE, Aldermaston, U.K. (1967) (unpublished).

12. J. C. Martin, "Nanosecond Pulse Techniques", AWRE, Aldermaston, U.K., Note SSWA/ JCM/704/49 (1970) (same as Ref. 14, Ch. 3).

13. V. S. Komelkov and G. N. Aretov, "Poluchenie Bolshykh Impulsnykh Tokov" ("Generation of Large Pulsed Currents"), *Doklady AN SSSR*, 110 (4), 559–561 (1956). (In Russian)

14. P. J. Rogers and H. R. Whittle, Fourth Symposium on Engineering Problems in Thermonuclear Research, Frascatti, Italy (1966) (unpublished; see Ref. 6).

15. P. J. Rogers and H. R. Whittle, "Electromagnetically Actuated Fast Closing Switch Using Polyethelene as Main Dielectric," *Proc. IEEE* 116, 173–179 (1969).

16. T. E. James, K. Harries, and R. O. Medford, "Development of Fast 100kV 1MA Solid Dielectric Switches and Associated Triggering Studies," *Proceedings of Symposium on Engineering Problems in Controlled Thermonuclear Research,* Livermore, CA, Univ. of California Report, CONF 650512, p. 77 (1965).

17. T. E. James, "Fast High Current Switching Systems for Megajoule Capacitor Banks," Culham Laboratory (Abington, Berkshire, U.K.), Report CLM-23 (1973).

18. P. Dokopuolos, M. Lochter, F. Woelbroeck, *Proceedings of Seventh Symposium on Fusion Technology,* Grenoble, published by Commission of European Communities, Center for Documentation and Information, Luxemburg (1972).

19. P. Dokopoulos, "A 2MA Metal-to-Metal Switch with 0.1 Microsecond Jitter Time," *Fifth Symposium on Fusion Technology,* Oxford, Paper 25 (1968). (Unpublished)

20. P. Dokopoulos, "Fast Metal-to-Metal Switch with 0.1 μsec Jitter Time," *Rev. Sci. Inst.* 39, 697 (1968).

21. R. S. Dike and R. W. Kewish, Jr., "The Development of a High Explosive Driven Crowbar Switch," *Proceedings of the Fifth Symposium on Engineering Problems of Fusion Research,* Princeton Univ., IEEE Pub. 73CH0843-3-NPS, p. 658 (1973).

22. D. P. Murphy, M. Raleigh, E. Laikin, J. R. Greig, and R. E. Pechacek, "Electron Beam Transport Through Atmosphere in Reduced Density, Current Carrying Channels," *Proceedings of the Ninth Symposium on Engineering Problems in Fusion Research,* Chicago, IL, IEEE Cat. No. 81CH1715-2 NPS, Vol. II (1981).

23. Detonator triggering requires electrical pulses at 1 kA, 1 kV level with a risetime of less then 1 μs. A commercially available unit that provides this output specifically for detonator firing, operating on a battery and triggered optically, is available from Reynolds Industries, Inc. (see p. 106 for full address).

24. J. C. Martin, U.K. Patent 988,777 (1970) (quoted in Ref. 7).

25. J. C. Martin and I. D. Smith, U.K. Patent 1,080,131 (1970) (quoted in Ref. 7).

26. Much of the detailed material on the switch design, including the discussion of the triggering circuit, is given in an undated internal Naval Research Note, "A Fast-Risetime Excitation System for the Production of Vacuum Ultraviolet Laser Emission," by J. D. Shipman, Jr., and R. W. Waynant. Ref. 1 above provides a brief description of this switch.

6
Liquid Dielectric Switches

INTRODUCTION

Switches using spark gaps submerged in insulating liquids perform on the same principles as spark gaps in gases. Their operation, the discharge through the liquid, depends on the breakdown properties of the insulating medium. Distribution of electric fields, leading to breakdown, is determined in a manner similar to that used for spark gaps, except that the dielectric constant, κ, unlike that for gaseous insulation, differs substantially from that for a vacuum (κ_0) and, naturally, must be accounted for when electric fields are calculated. In some liquids, such as water, the value of the relative dielectric constant is rather high, i.e., $\epsilon = \kappa/\kappa_0 = 80$. Furthermore, in the design of the switches based on liquid insulation in the gap, greater mechanical stresses can arise. Thus, greater care must be exercised in areas of contact between electrodes and the structural parts of the switch as compared with that already discussed for gases in Chapter 3.

The behavior of liquid dielectric switches is described in this chapter in the same terms as other spark gap switches. The amount of variation in switching voltage, as well as jitter in the switching time relative to some feature in the waveform of the voltage applied across the gap (or jitter in the delay relative to the trigger pulse) discussed earlier, both determine the performance of the liquid dielectric switches.

The main motivation for the development of liquid dielectric switches is their mechanical or structural compatability with the high power pulse systems which use high breakdown strength liquids to store electrical energy before delivery to the load. Such switches have been used in very large pulse generation in facilities such as AURORA[1], which uses oil, and in BLACK-JACK V[2] and GAMBLE II[3] and others[4], which use water for a dielectric. By using the same medium for switching as for energy storage, it is possible for many designs to eliminate critical interfaces such as diaphragms for separating liquid from the gas in the switch. Additionally, the high breakdown value of liquid dielectrics, in the range of 100 to 1000 kV/cm is an order of magnitude higher than even for high pressure gases. Therefore, the resulting inductance of the switch arc channel can be reduced substantially, making such switches especially attractive for use in low impedance generation[2, 3, 4].

The development of liquid switches for high power applications has been limited to custom designs. One of the difficulties frequently encountered in the design is the effect of the pressure pulse generated by the discharge channels in the nearly incompressible liquids. For example, it was necessary to remove the switch from the vicinity of the load (an electron beam diode, using relatively fragile water-vacuum interface) in the GAMBLE II generator[3], as noted in Chapter 2. Because of the large breakdown strength, it is difficult to use such switches at voltages below 50 kV, because the small gap between the switch electrodes cannot be set with sufficient accuracy to provide predictable switching. Water is self-healing. Oil, on the other hand, is carbonized during switching, requiring some means of removal or dispersion of the carbon to maintain reproducible operation.

Triggering of the liquid dielectric switches is an important part of the design for high power applications. It is of primary importance in synchronization of multiple pulser systems which are combined to drive single loads[5] or individual loads which must be operated simultaneously[6]. Although the basic principles of triggering are similar to those already discussed, for example, in Chapter 5, triggering becomes more difficult to achieve because of the need to handle mechanical impulses, and, for the case of high dielectric liquids, because of greater loading of the trigger generator. Operation of a triggered switch is described later in this chapter, with a demonstration of triggering techniques useful in designing liquid dielectric switches.

Triggering of the liquid dielectric switches can be used to provide multichanneling of the discharge between the switch electrodes. This helps to reduce inductance even further in comparison with that of gas switches. It further helps to distribute pressure pulse and erosion effects over a larger area of the electrodes to help prevent structural damage to the storage system and the load envelope.

Use of liquid dielectric switches is more recent than that of gas spark gaps and there is much less reporting at least easily accessible, that describes the switch design, performance and operational data. For example, V. S. Komelkov's otherwise excellent pulse power design text[7] has no mention of liquid dielectric switches.

ELECTRICAL CHARACTERISTICS OF LIQUIDS

General Characteristics

There is a large variety of liquids that can serve as insulators in the inter-electrode region of a spark gap switch. It makes practical sense to use only a small selection of liquids. Insulating "transformer" oil represents a type of liquid with large dielectric strength and small dielectric constant, making it preferable for low inductance switching. Because, typically, hydrocarbons have resistivity in a 10^{15}–10^{20} Ω-cm range, they can be used in switches which must hold-off voltages for long periods. Water has a very different behavior. When purified, its resistivity is 2 to 3 × 10^7 Ω-cm. In experimental pulse systems using water for insulation, such as GAMBLE II generator[3], water resistance is maintained at about a 5 × 10^6 Ω-cm level by the use of commercial ion-exchange demineralizing units. Relatively low resistivity accounts for the use of water switches in holding off only short duration (< few μs) electric fields. Indeed, water dielectric pulse lines are charged very rapidly for two reasons: their dielectric strength increases with the decrease in the charging time (as discussed below) and the resistive losses caused by low resistance would be too high in practice for the charging supplies to overcome for charging times exceeding 100 μs. The relative dielectric constant of water is high (ϵ = 80), suggesting that the high capacity associated with the electrodes must be accounted for in the pulser designs using water breakdown switches. One well known and troublesome effect because of water's dielectric constant being high, is the appearance of the so-called "prepulse," i.e., lack of complete isolation of the load from the voltage being applied to the pulser before switching. (Prepulse associated with water dielectric switches is discussed in Ref. 6.)

Recently, pulsed high voltage dielectric properties of an ethylene glycol/water mixture (approximately 50% water) were studied as a potential insulator for high power pulsed generation[8]. It is a liquid with properties somewhere between oil and water. Its dielectric constant can approach that of water (at near freezing point) and its resistivity, ρ, is some orders of magnitude closer to that of oil. Thus, its intrinsic time constant, in seconds,

$$\tau = \epsilon \kappa_0 \rho \qquad (6\text{-}1)$$

(where κ_0 = 8.85 × 10^{-12} F/m and ρ is in ohm-meters) has a value in the millisecond regime. The observed high voltage behavior, in small scale tests, (electrode areas of less than 100 cm^2 separated by a few mm) shows the breakdown strength to be much higher (~150 kV/cm) in the millisecond regime than would be expected in pure water, if the scaling given by Eq. 6-2, below, is valid in that pulse duration range. The ability to sustain such high fields in a medium with a high dielectric constant for millisecond time scales suggests that the use of such mixtures in future compact pulse power systems, driven by inertial-inductive generators (such as self-excited homopolar generators[9] or repetitive output "Compulsator" rotational machinery[10]). This ability indicates that the ethylene glycol/water mixtures may be used in spark gap switches, also.

Breakdown Strength

Detailed breakdown characteristics are now considered for oil and water only. Breakdown in these liquids has been studied under various conditions, as defined by applied electric stress, duration of applied voltage, and impurities and gases in the liquids.

J. C. Martin[11] has developed breakdown scaling relations for a wide range of voltages and pulse times. He gives, for uniform fields, the breakdown value, F (in MV/cm), in terms of the duration of applied voltage, t_{eff}, in microseconds, and the electrode area, A (in cm^2):

$$Ft_{eff}^{1/3} A^{1/10} = k \qquad (6-2)$$

The breakdown dependence on electrode area A is in contrast to the breakdown dependence on volume, exhibited by solids (for example, Eq. 5-1, in Chapter 5). This suggests that the breakdown is associated with phenomena related to the conditions on the surface of electrodes rather than to the weak points in the dielectric volume. The value of k for transformer oil is approximately $k = 0.5$ and $k = 0.3$ for water. Experiments with divergent geometries which unbalance the fields on the electrodes (e.g., in a point-plane geometry) show, however, that water has considerable polarity effect. The above value of k corresponds to positive breakdown streamers. If the field at the negative electrode is increased sufficiently above that of the positive electrode, so that breakdown streamers start from the negative electrode, then the value of $k \doteq k^- = 0.6$ for water. Oil has similar polarity effect. Smith[12] finds $k = 0.5$ for t_{eff} ranging from 0.3 to 1.0 μs.

Figure 6-1 is a plot of Eq. 6-2 over four orders of change in area[11] (defined as that area over which field strength is 90% or more of the maximum) for pulse times in the microsecond regime. This plot has been generalized (1) for all electrode shapes by introducing a factor α which accounts for the departure of the electrode surface field, F_{max}, from the mean field, F_{mean}, between the electrodes (with α not exceeding a value of 1.2 for most electrode configurations), and (2) for any monotonically rising pulse shape

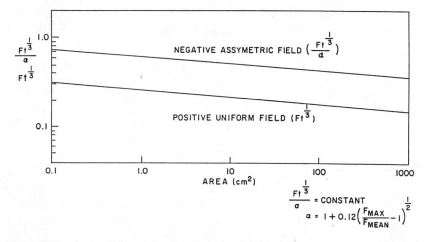

Fig. 6-1. Impulse breakdown of water. Variable t is equivalent to t_{eff} in Eq. 6-2. Adapted from P. D. A. Champney, "Impulse Breakdown of Deionized Water with Assymetric Fields," Note 7, *Pulsed Electrical Power, Dielectric Strength Notes, PEP-5-1,* Vol. 1, AFWL Report TR-73-167 (1973).

of duration t_{eff} (defined as the time during which the voltage applied to the electrodes is 63% or more of the breakdown value).

A. P. Alkhimov[14] et al., pursuing the theory that breakdown in water is initiated by air bubbles on the surface, found that the increase in water pressure (and consequent reduction in the bubble size) leads to a significant increase in breakdown strength. Assuming that their electrode areas were of the order of few cm², the breakdown value of 700 kV/cm they obtained for a pulse duration of 10 to 15 μs at a pressure of 150 atm exceeds the normal breakdown value, given by Eq. 6-2, by about a factor of 5. These authors also agree reasonably with the time dependence of breakdown as expressed in Eq. 6-2.

Based on similar considerations of breakdown initiation as the above authors (Ref. 14) suggested, V. V. Vorobiev et al.[15] found another method for increasing the breakdown strength of water. By diffusing a salt solution through an electrode grid into the main inter-electrode region, they were able to smooth out the field irregularities on the surfaces by gradual distribution of high conductivity over a region of approximately 1 mm in thickness adjoining the surface. The result was an increase in the breakdown strength by about a factor of 3. While this may not be a practical method for increasing the dielectric strength in high energy storage pulse lines, it appears useful for small switches requiring low inductance (i.e., short gaps) since the configuration of such "diffuse" electrodes is stable for a period of hours. However, because the discharge would disturb the appropriate distribution of conductivity, flushing of the switch would be required.

Finally, insulating oil and other liquids with high resistivity and liquids with intermediate resistivity such as water can, under conditions of a strong applied field, undergo a redistribution of charge in the body of the liquid. Such motion of the charge, driven by the applied field, corresponds to the motion of ions[16]. Typical redistribution times are in the millisecond range. Selecting proper electrode materials, Zahn[17] has shown that it is possible to increase breakdown strength by about 30% as a result of charge redistribution. In the case of steel and aluminum electrodes operated in the polarity that produces charge injection, breakdown is at 140 kV/cm. Reversing the polarity of the voltage applied to the electrodes reduces this breakdown strength to 100 kV/cm.

Channel Formation and Its Characteristics

Breakdown of liquids is characterized by distinct luminous channels. Conditions preceeding the formation of channels, their formation, and the character of the channels has been studied extensively. For example, Alkhimov et al.[14] have developed a model of streamer formation through the use of low current discharges to study breakdown at field strengths of 100 to 500 kV/cm. As the field increases and time to breakdown becomes shorter, the initial "free" electrons are accelerated to a point where they reach sufficient energy to ionize. The number of free electrons in the liquid volume is thus multiplied with the simultaneous dissociation of water, forming free hydrogen. The hydrogen bubble then leads to formation of the luminous channel.

J. C. Martin[18] with others at the Aldermaston Weapon Research Establishment in United Kingdom has studied high current streamer initiation and subsequent discharge properties. He has suggested semi-empirical formulas for the resistive and inductive phases of the discharge streamer as well as for the propagation velocities of the streamer. It is these relations which make it possible to predict the performance of liquid dielectric switches and has been used in the design of switches discussed below.

Streamer velocities behave quite differently in various liquids and change their propagation velocities in different voltage regimes. The streamer velocity is given in Ref. 11 for point (or edge)-to-plane electrode geometry in the 100 to 1000 kV range for a few organic liquids:

$$U_{av} = d/t = kV^n \qquad (6\text{--}3)$$

Here U_{av} is in cm/μs and V is in MV. The values of k and n are:

	k^+	n^+	k^-	n^-
Oil[11,13]	90	1.75	31	1.28
Carbon tetrachloride[11,13]	168	1.63	166	1.71
Glycerine[11,13]	41	0.55	51	1.25
Water[13]	9	0.60	16	1.09

where + and − indicate the initiation from positive and negative electrodes respectively.

For voltage in 1–5 MV range, both polarities in oil obey the following relationship:

$$U_{av}d^{1/4} = 80V^{1.6} \qquad (6\text{–}4)$$

The streamer velocity behaves differently for water. The best fit, according to J. C. Martin[11] is (again, in cm/μs for V in MV and in μs for t):

$$U_{av}t^{1/2} = 8.8V^{0.6} \quad \text{for positive electrode} \qquad (6\text{–}5a)$$

$$U_{av}t^{1/2} = 16V^{1.1} \quad \text{for negative electrodes} \qquad (6\text{–}5b)$$

(Reference 6 provides a measurement of instantenous streamer velocity in water as a function of the mean electric field remaining in the gap untraversed by streamers.)

The resistive phase, i.e., the resistive risetime of the discharge current, for example, following the discussion in Ref. 4 (P. VanDevender) is

$$\tau_R = \frac{230\rho^{1/2}}{(NZ)^{1/3} E^{4/3}} \text{ seconds} \qquad (6\text{–}6)$$

where E is the average electric field at breakdown in V/m, Z is the impedance of the circuit driving the discharge (i.e., the characteristic impedance of the electrode and feed plates to a distance, equivalent in terms of a transit time, equal to the switch risetime). N is the number of discharge channels and ρ is the *specific* gravity of the dielectric. Equation 6–6 is an expression generalized to multiple channels, also discussed in Chapter 5 (i.e., Eq. 5–4). Similarly the inductive phase, i.e. that part of current risetime which is limited by the inductance of the discharge channel, L, is for N channels:

$$\tau_L = L/NZ \qquad (6\text{–}7)$$

J. C. Martin estimates the channel inductance by assuming the expansion of the channel to be about 10^6 cm/s in liquids (Ref. 4). An adequate approximation for the inductance during the risetime (when energy is still being deposited) is to use the expression for the inductance of the wire (in nH) with radius a, and length ℓ (in cm) fed by a disc of radius b, so that

$$L_{channel} = 2 \cdot \ell \ln (b/a) \qquad (6\text{--}8)$$

For b of few centimeters and 10 ns risetime (i.e., a $\approx 10^{-2}$ cm), the logarithmic term is about 6 to 7, and inductance is 12–14 nH/cm.

To estimate the amount of energy that is deposited in a channel, Ref. 18 uses the following expression for ohmic dissipation during the resistive part of the risetime:

$$W_{loss} = (V_0^2/4Z) \, \tau_R \qquad (6\text{--}9)$$

with V_0 being the initial gap voltage. VanDevender finds other authors, R. C. O'Rourke[19] and O. Milton[20], agree with this approach with minor corrections, if the gap length, inductance, and capacity are not too great. VanDevender[4] has shown that measurements of dissipated energy in such gaps agree with Eq. 6–9 over a wide range of parameters. He further discusses the partition of energy absorbed by the switch. Most of this energy goes into ionization, excitation, and heating of the channel and a fraction of it is expended as acoustic energy in a shock wave. It is important to know the characteristics of the shock in determining the mechanical forces on the design of switch parts to assure that they can withstand repeated switching. The energy contained in the pressure pulse is given[4] (following the scaling of explosion hydrodynamics discussed by R. H. Cole[21]):

$$W_p = \int d\omega r^2 \int_0^\infty (P^2/\rho U_s) \, dt \qquad (6\text{--}10)$$

where $P(t)$ is pressure in pascals at distance r in meters from the channel, ρ is the initial water density, U_s is the shock velocity (approximately equal to sound speed in many cases), and $d\omega$ and dt are the differentials of the solid angle and time respectively.

To obtain some measure of switch pressure characteristics, Fig. 6–2 is reproduced from Ref. 4, showing that the pressure pulse reaches more then 10^7 Pa (approximately 100 atm). This pressure can be one or two orders of magnitude higher at the point of contact of the streamer with the electrode, as can be inferred from the yield strength of the electrode materials. The calculated acoustic energy is given as 8% (for $\tau_R = 18$ ns) of the total dissipation of 1.5 kJ in the channel.

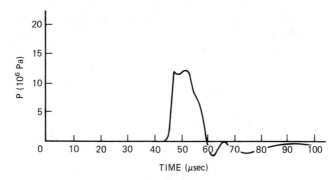

Fig. 6-2. Pressure profile caused by a discharge in water at a distance of 21 cm from the 7.6 cm long gap. Reproduced from J. P. VanDevender, Ref. 4.

The highly concentrated channel formed in liquids also leads to some electrode damage. Investigations of the electrode erosion made of several different materials are given in Ref. 22 and 23. Burton et al.[24] measured the erosion of steel rods of 1.6 mm diameters at much higher energies (up to 30kJ/channel) relevant to high power switching. Short circuit discharge of the GAMBLE II pulseline, using a 5.8 Ω internal impedance (and output pulse duration of 50 ns) configuration, charged up to 4 MV, produces current levels of up to 0.75 MA in the switch. Resulting electrode erosion was measured to be 1.5 to 2.0 mm per discharge. These results represent an extreme case, since switching into a finite resistance load would lower the amount of charge transfer through the switch by a factor of 10 or more. Finally, the electrode erosion also indicates that the rate of material removal at the point where streamers start is about 10 times that at the opposite electrode.

Causes of Breakdown at Liquid-Solid Interfaces

Finally, the breakdown properties along solid-liquid surfaces, considered in the design of liquid dielectric pulse lines, round out the liquid dielectric breakdown characterization needed for the switch design. This is similar to the interface problems already discussed in Chapter 3 and again in Chapter 5 in conjunction with gas and solid dielectric switches. A few studies that are relevant to water dielectrics, for example, by J. K. Burton[25] and A. R. Miller [26], have concluded that the voltage across the interface surface must be the same (or higher) than that required to break down the dielectric, provided no air bubbles are found on the surface and that the triple-point (i.e., the juncture of the solid-liquid surface with the metallic electrode surface) is clean of debris and air bubbles. Reference 26 shows that more then 150 kV/cm can be maintained across large diaphragm surfaces in water pulse lines, assuming that voids and other imperfections in the diaphragm itself are avoided. In general, when enhancement of the electric field on the surface is avoided, the breakdown at the interface is at least as high as within the liquid dielectric[26].

In applying the breakdown criteria to liquid gap switches, as for example in the switch (shown in Fig. 2-2 in Chapter 2) which uses an acrylic-water interface (as well as an acrylic-SF_6 interface), there are three failure modes that should be considered. These breakdown modes involving the interface have been classified by A. R. Miller[27] as follows:

1. Breakdown streamers begin at points of field enhancement on the conductors and grow until they intercept the plastic interface from which point they then travel along the surface until they reach the opposite electrode, shown in Fig. 6-3A. The field enhancement which starts the streamer may be due to an excessively small radius of curvature of the conductor surface or to particulate contamination of the electrode surface.

2. Streamers begin at points of field enhancement on the plastic surface (i.e., dirt, bubbles, deep scratches, etc.) away from the conductors, usually growing in both directions until they reach the conductors, as shown in Fig. 6-3B.

3. Streamers originate in the plastic in impurities and grow through the bulk material, usually emerging at a surface and then traveling to the conductor or growing through an imperfection in the bulk of the plastic (i.e., a seam, crack, or low density area) until reaching a conductor, as shown in Fig. 6-3C.

Thus the breakdowns are caused by enhancement due to contaminants in the water and in, or on, the plastic, or to sharp edge field enhancement on the conductors. The breakdown process continues over many charging cycles, or it may be prompt. The foregoing constitutes the less controllable factors to be considered in assessing the viability of a particular use of interface, and relate, importantly, to the operating environment. The parameters of mechanical and electrical operating stress are controllable design factors and must also be observed.

HIGH PERFORMANCE SWITCH DESIGN

The breakdown properties of liquids (including the liquid-solid interface regions), the electrical characteristics of the ensuing discharge channels, and

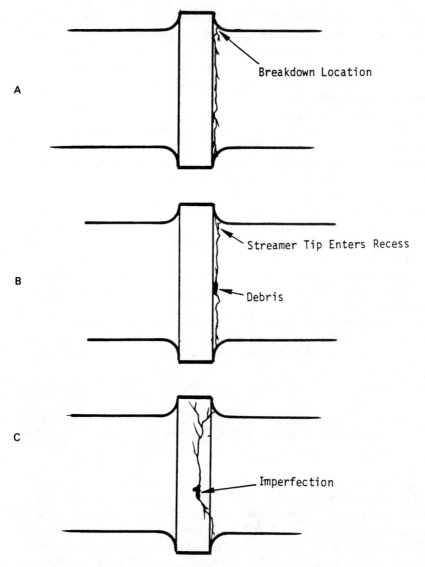

Fig. 6-3. Electrical breakdown modes associated with the liquid-solid interfaces. Reproduced from Ref. 27.

the effects of pressure impulses produced by the discharge, comprise the three crucial elements to be considered in the design of liquid dielectric switches. In contrast to solid dielectric switches (discussed in Chapter 5), liquid dielectric switches can be used for many discharges without replacement of any of their parts. For this reason, as well as their advantageous electrical properties, they are favored for those applications in which pulser configurations make difficult reaching the switches for routine operation or maintenance. Of the largest pulse generators, AURORA[1] employs four synchronized switches using oil and BLACKJACK V[2] uses water as the dielectric with its switches operating in multichannel mode.

This section is concerned with the development and design of untriggered and triggered *multi*channel liquid dielectric switches. Their performance as compared to *single* channel operation is much better, but multichannel design guidelines are much more complicated.

Untriggered Formation of Multiple Channels

The experiment described in Ref. 28 demonstrates simultaneous formation of three channels in an untriggered water switch. Fig. 6-4 shows, at bottom, the parallel plate electrodes with three field-enhancing rounded stubs, at negative potential. The electric field strength at the surface of the stubs is tripled relative to the average field between the electrodes, so that the breakdown streamer initiates at the negative electrode; this is predicted from Eq. 6-2, (or from Fig. 6-1 for an electrode area of about 1 cm^2 and $\alpha = 1.2$) for 200 kV applied across the gap with the voltage rise, $t_{\text{eff}} = 200$ to 850 ns, giving about 1.2 MV/cm; in Eq. 6.2, F is the maximum field. Thus the average field of -400 kV/cm across 0.5 cm gap leads to a breakdown voltage of 200 kV. In the experiment, this switch was operated near 250 kV with gap spacings varying between 0.50 cm and 0.61 cm (measured within 5%). The peak voltage varied from 185 kV to 255 kV and t_{eff} varied from 200 ns to 850 ns due to the wide variation of the breakdown point of the switch in the primary circuit of the pulse-charging transformer which applied the voltage to the switch. Of 23 tests, only in 3 cases did all three channels not establish some conducting path.

Figure 6-4 (first frame) shows a closing of two of the gaps while a streamer (bright channel that does not completely span the electrode gap) is seen to start on the middle gap. The start of a streamer at the negative electrode is a check on the correct choice of electrode curvature and electrode spacing.

To study the difference between channels of different brightness to correlate such differences with the distribution of currents in the channel, a streak camera was used to observe the time of initiation of luminescence of the channels. Fig. 6-5 shows four examples of the performance of a four-channel switch operating near 250 kV. Both time-integrated and streak photographs are shown. The number 4 test shows the simultaneous closing (within 10 ns resolving time) of the two adjacent channels. It should be

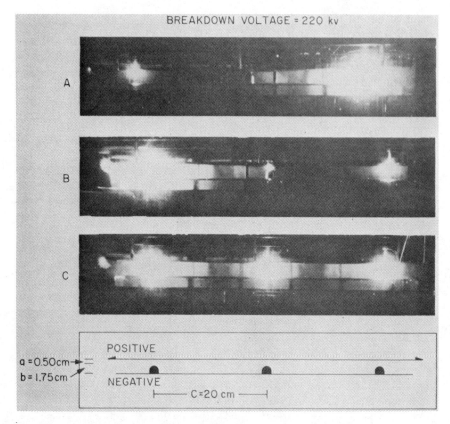

Fig. 6-4. Switch geometry and dimensions are shown at top. Frames A, B and C are time exposures of the switch closing, with successful multichannel operation seen in frame C.

noted that the time-integrated brightness is the same in both channels. Two tests (1 and 2) in Fig. 6–5 show that a second channel closing may occur as late as 150 ns after the first channel has closed and started to conduct. The amount of energy carried by the second channel is small, as indicated by its luminosity. The third test also shows that the streamer seen in one of the channels starts late (~ 100 ns) after the first channel closing. The time delay between channel closings approximates the half-period of the discharge. This indicates that channel breakdown occurs as a rseult of reversal (wrong polarity on field-enhanced electrode) with the resistive and inductive voltage across the nearby channel already established, being sufficient to break down the gap.

The representative data in Figs. 6–4 and 6–5 show that all three channels

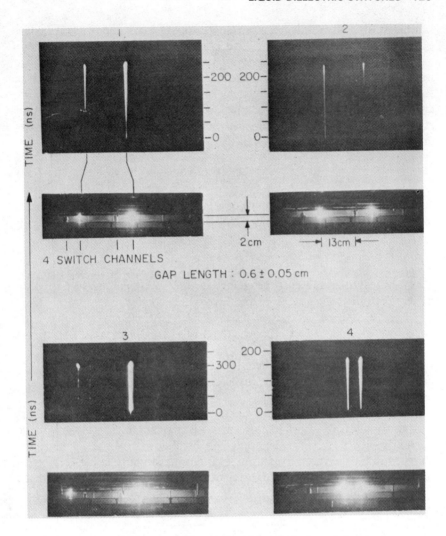

Fig. 6-5. Streak and time exposure photographs of the switch channel closing.

can be made to close if the channel-to-channel separation, ℓ, is sufficiently large. This separation is related to the communication time between channels.

$$T = \frac{2c}{\sqrt{\epsilon}\,\ell} \tag{6-11}$$

where ϵ = the relative dielectric constant of water and c = the velocity of light. Because the voltage across the breakdown channel does not drop instantaneously, but is reduced according to the resistive and inductive components of the switching time, τ_R and τ_L respectively, the voltage difference across *neighboring* gaps will be attenuated in time $T + \tau_R + \tau_L$. Assuming, initially, no deviations of breakdown strength from the curve of Fig. 6–1, the only cause for preferential closing of a given channel would be the increased electric field on the channel whose electrodes are spaced more closely. The inaccuracy in spacing, δ, that can be tolerated is therefore governed by the data of Fig. 6–1. Thus, rewriting Eq. 6–2:

$$F_{max}^- = kt_{eff}^{-1/3}, \quad k = \text{constant} \qquad (6\text{–}12)$$

By differentiating this equation to obtain a change in the breakdown field, dF_{max}^-, as the gap ℓ changes by a small amount dx, then

$$\frac{dF_{max}^-}{dx} = \tfrac{1}{3} \, t_{eff}^{-1} F_{max}^- \frac{dt}{dx} \qquad (6\text{–}13)$$

The factor dt/dx represents the change in time to breakdown as ℓ changes by dx. Thus $dt = T + \tau_R + \tau_L$ and Eq. 6–13 becomes:

$$\delta = \frac{dF_{max}^-}{F_{max}^-} = \tfrac{1}{3} \, (t_{eff}^{-1}) \, (T + \tau_R \, \tau_L) \qquad (6\text{–}14)$$

In the experiments described here (where the switch was driven by a 2.2 Ω line) the computed value of δ was less than 1% ($t_{eff} \simeq 500$ ns, $\tau_R \simeq 2.5$ ns/channel, $\tau_L \simeq 1.0$ ns/channel*, $T \simeq 6$ ns).

Additionally, the spread of breakdown data will influence the channel closing simultaneity. This imposes a risetime requirement on the voltage applied to the gap, since the spread in breakdown voltage ∇V must be spanned in a time shorter than the signal propagation time between channels, $T + \tau_R + \tau_L$. If dV/dt represents the voltage rise across the switch, then the condition

$$\frac{dV}{dt} \geq \frac{\nabla V}{T + \tau_R + \tau_L} \qquad (6\text{–}15)$$

must hold to allow the neighboring channels to close.

Since no preference for a given gap closing has been observed, the statistics of simultaneous breakdown in the epxeriment described here is dominated by the intrinsic value of the scatter in the breakdown voltage. Indeed the observed mean value, ∇V, of 9% indicates that dV/dt is too small to

---

*τ_R is estimated from Eq. 6–6; τ_L is estimated from Eq. 6–7, where the inductance, L, is calculated by assuming a channel diameter of 10^{-2} cm, according to J. C. Martin's work[11].

satisfy the inequality 6-15 or that the gap is too wide due to low enhancement. J. C. Martin[4] indicates that a somewhat lower deviation (7%) was observed in the data plotted in Fig. 6-1. (This is significantly better than the 12% deviation found for oil and polyethelene). It is therefore expected on this basis that a typical 1 MV pulser would require 3×10^{12} V/s charging rate for simultaneous switching of channels 20 cm apart in a switch with main plate separation ($a + b$ of Fig. 6-4) being 2.5 cm if 50 Ω/channel generators are used.

Triggering of Multiple Channels

To add precise switching time, such as would be needed to synchronize high power pulses of 50–100 ns duration, a triggered water multichannel switch has been designed by J. D. Shipman of the Naval Research Laboratory; it has been tested and operated on the terawatt generator, GAMBLE II[3]. This switch replaces a self-closing switch operated by overvoltaging the switch gap. The new switch, placed in the position occupied by the old switch between the pulse-forming line and the coaxial transformer of GAMBLE II, has been designed and fabricated as shown in Fig. 2-2 of Chapter 2. The switch operates as follows:

During the charge of the pulse forming line to V volts, the disc-shaped trigger electrode, the $CuSO_4$ resistor, and the large negative electrode of the SF_6 trigatron switch will charge up to $-V/6$ volts by capacitive division. The potential conditions at this time are shown in Figure 6-6a. Note that at this time very little field enhancement exists at the edge of the trigger disc. When GAMBLE II is operating at the maximum level of 100 kJ in the pulse-forming line, the voltage, V, is 6.7 MV. The acrylic–SF_6 interface has been designed so that the voltage distribution along it is uniform within about 20% to minimize the switch dimensions.

The next step in the operation of the switch is for the SF_6 trigatron switch to receive a positive trigger signal of about 100 kV (which could easily be derived from the firing of the intermediate store output switch). This signal is picked up by a spherical electrode inserted in the water above the intermediate store center conductor. It is transmitted through an RG-19/U cable with an appropriate delay length to the trigger electrode of the SF_6 switch.

Closure of the trigatron switch drops the trigger disc potential to that of the nearby positive electrode in about 20 ns. The $CuSO_4$ resistor ($\sim 30 \, \Omega$) is near the critical damping value for the LC circuit formed by the capacitance of the trigger disc assembly (0.46 nF) and the inductance of the SF_6 coaxial switch assembly (~ 90 nH), and no voltage overshoot occurs. As the trigger disc drops in potential, the field enhancement at its edge will start streamers from the edge toward the negative electrode on the end of

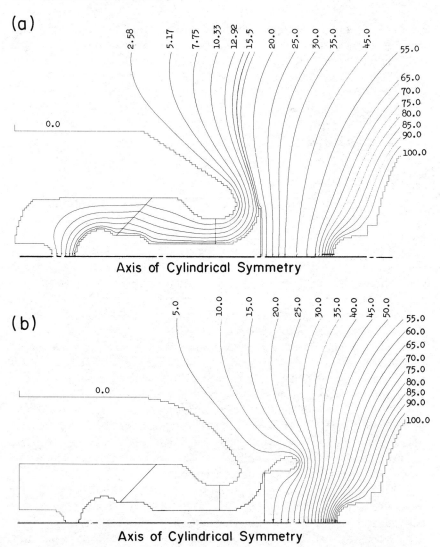

Fig. 6-6. Potential distribution of the switch shown in Fig. 2-2 in Chapter 2: (a) distribution prior to any triggering, (b) distribution after the potential of the trigger disc has been lowered to that of nearby electrode and streamers have completed part of the way to the far electrode, and (c) distribution after the streamers have closed and the trigger disc potential equals that of the far electrode.

the pulse-forming line. The transit time for these streamers will be 30 to 40 ns. Assuming that several streamers start around the circumference of the disc, Figure 6-6b is an approximation to the potential distribution when the streamers are part way across.

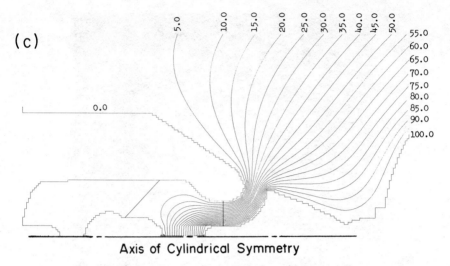

(c)

Fig. 6-6. *(cont.)*

When the streamers do reach the pulse line, the potential of the trigger disc will rise rapidly toward the high negative potential of the pulse line. The 30 Ω CuSO$_4$ resistor performs the essential function here of decoupling the trigger disc from the low inductance switch connection to the positive electrode. Figure 6-6c shows what the potential distribution would look like if the potential of the trigger disc rose all the way to the potential of the negative electrode. Actually, before this potential is reached, streamers already have started from the left edge of the trigger disc and closed to the positive electrode with a transit time of 10 to 20 ns. These streamers close before the acrylic-water interface suffers breakdown due to interface streamers.

This switch has been fabricated and operated at the full GAMBLE II power output of more than 10^{12} W. It is shown in operation in Fig. 6-7, consistently operating in multichannel mode.

SCALING

The scalings describing various phenomena, which determine switch performance, for example, the breakdown streamer velocity described by the scaling relation embodied in Eq. 6-3, as well as operating conditions of switches (e.g., how they are incorporated into a pulser) based on such phenomena, must be considered carefully in designing liquid switches. It has already been noted that the time dependence in relations for breakdown (Eq. 6-2) has been established for charging times (t_{eff}) limited to 1 μs. This dependence becomes weaker as charging time increases. It is also weaker

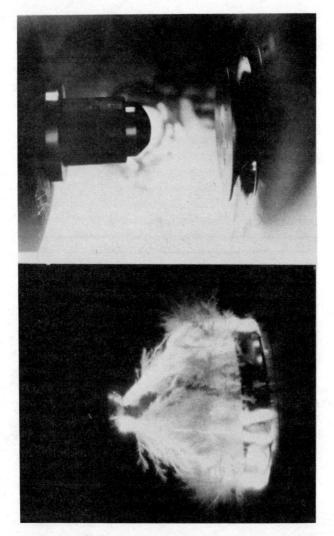

Fig. 6-7. Triggered water switch installed in GAMBLE II. The upper left-hand frame shows the switch electrodes; the cylindrical variable length gap is at left; and the trigger disc electrode is mounted on the "ground" electrode at the right. Other frames, depicting three separate discharges, show two sets of streamer configurations during switching, those spanning "ground"-trigger gap, and trigger-high voltage gap.

$(t^{1/5}_{eff})$ relative to Eq. 6-2 for higher water pressures[29]. Similarly, the effect of pressure on the breakdown voltage disappears[2] as the electrode area increases to values substantially larger than those normally associated with switch electrodes.

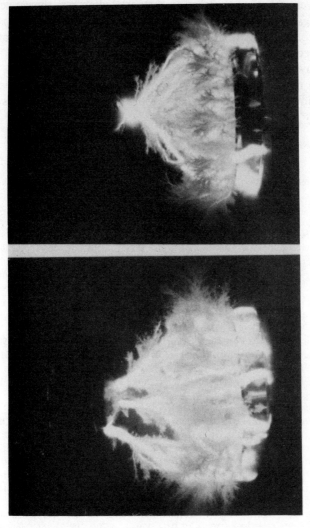

Fig. 6-7. (*cont.*)

The validity of relations such as Eq. 6-9, while providing an estimate of the energy deposited during the time τ_R (which measures the resistive phase of the switch), also must be re-examined when applied in a regime other than that for which the measurements were made. For example, for very high voltage switches, the value of Z in Eq. 6-9 may not be that of the pulse line, but some effective value associated with the geometry of the current feed to the streamer channel. Finally, large separation of high voltage switch electrodes means that a substantial change in the capacity of the switching gap[6] has occurred, suggesting that a current through a trigger plate resistor provides an ohmic drop which reduces the trigger plate voltage and so affects triggering.

High power liquid dielectric switches, especially those based on water insulation, remain to this day developmental, and further engineering data and techniques needed to provide complete design prescriptions continue to evolve.

REFERENCES

1. B. Bernstein, I. Smith, "AURORA, an Electron Accelerator" *IEEE Trans. Nucl. Sci.* **20,** 294 (June, 1973).
2. A. R. Miller, "Power Flow Enhancement in the Blackjack 5 Pulser," *Digest of Technical Papers of the Fourth IEEE Pulsed Power Conference,* Albuquerque, N.M., IEEE Cat. No. 83CH1908-3, pp. 594–597 (1983).
3. L. S. Levine and I. M. Vitkovitsky, "Pulse Power Technology for Controlled Thermonuclear Fusion," *IEEE Trans. on Nucl. Sci.* **NS-18,** 255–264 (1971).
4. A variety of smaller water-insulated pulse generators have been built, starting with that of J. C. Martin, Cornell University Seminar on High Voltage Technology (1966) (unpublished); a representative pulser is that used for the water dielectric switching studies (J. P. VanDevender, "The Resistive Phase of a High Voltage Water Spark," *J. Appl. Phys.* **49,** 2617 [1978]).
5. T. H. Martin, "Pulsed Power for Fusion," *Digest of Technical Papers of the Second IEEE International Pulse Power Conference,* Lubbock, TX, IEEE Cat. No. 79CH1505-7, pp. 2–8 (1979).
6. G. M. Wilkinson, "Time Dependent Capacitance Effects in Water Dielectric Switching," op. cit. Ref. 2, pp. 323–326.
7. V. S. Komelkov (see Ref. 4, Ch. 3).
8. D. B. Fenneman, "Pulsed High-Voltage Dielectric Properties of Ethylene Glycol/Water Mixtures," *J. Appl. Phys.* **53,** 8961 (1982).
9. R. D. Ford, D. Jenkins, W. H. Lupton, I. M. Vitkovitsky, "Operation of Multi-Megajoule Inertial-Inductive Pulser," *Digest of Technical Papers of Third IEEE International Pulsed Power Conference,* Albuquerque, NM, IEEE Cat. No. 81CH1662-6, pp. 116–121 (1981).
10. W. L. Bird, W. F. Weldon, "Design of a Compact, Lightweight Pulsed Homopolar Generator Power Supply," op. cit. Ref. 9, pp. 134–141.
11. J. C. Martin, "Nanosecond Pulse Techniques," *Pulsed Electrical Power Circuit and Electromagnetic System Design Notes PEP4-1,* Vol. 1, AFWL-TR 73-166, Note 4 (1973).
12. I. Smith, "Short Pulse Insulation and Breakdown: A Phenomenological Review," *Proceedings of the Conference on Electrical Insulation and Dielectric Phenomena,* Pocono, PA (1979), (unpublished).
13. G. Herbert, AWRE Note SSWA/HGH/6610/104 (1966) (unpublished).
14. A. P. Alkhimov, V. V. Vorobyev, V. F. Klimkin, A. G. Ponomarenko, and R. I. Soloukhin, "The Development of Electrical Discharge in Water," *Sov. Phys. Doklady* **15,** 959–961 (1971).
15. V. V. Vorobyev, V. A. Kapitonov, and E. P. Kruglyakov, "Increase of Dielectric Strength of Water in a System with 'Diffussion' Electrodes," *ZhETF Pis. Red.* **19,** 95–98 (1974) (in Russian).

16. M. Zahn and H. Chatelon, "Charge Injection Between Concentric Cylindrical Electrodes," *J. Appl. Phys.* **48**, 1797–1805 (1977).
17. M. Zahn, Y. Oaki, K. Rhoads, M. LaGasse, and H. Matsuzawa, "Electro-Optic Charge Injection and Transport Measurements in Highly Purified Water and Water/Glycol Mixtures," *IEEE Trans. on Electrical Insulation* **EI-20** (1985).
18. J. C. Martin, AWRE Note SSWA/JCM/1065/25 (1965) (unpublished).
19. R. C. O'Rourke, Maxwell Laboratories Report MLI-5-69 (1969) (unpublished).
20. O. Milton, Sandia Laboratories Report, SAND-76 0086 (1976) (unpublished).
21. R. H. Cole, "Underwater Explosions," Princeton University Press, Princeton, NJ (1948).
22. V. E. Illin and S. V. Lebedev, "Destruction of Electrodes by Electric Discharges of High Current Density," *Sov. Phys. Tech. Phys.* **7**, 717 (1963).
23. M. Motoki, C. Lee, T. Tanimura, "Electrode Erosion Due to Transient arc Discharge in Dielectric Liquid," *Electrical Engineering in Japan* **87**, 75–83 (April 1967). This reference discusses the arc machining of metals and deals with erosion resulting from *repetitive* pulsing at low power.
24. J. K. Burton, D. Conte, W. H. Lupton, J. D. Shipman, Jr., and I. M. Vitkovitsky, "Multiple Channel Switching in Water Dielectric Pulse Generators," *IEEE Proceedings of the Fifth Symposium on Engineering Problems of Fusion Research Conference,* Record Cat. No. 73 CH0843-3 NPS, pp. 679–683 (1973).
25. J. K. Burton, Private communication (1970).
26. A. R. Miller, "Sub-Ohm Coaxial Pulse Generators, Blackjack 3, 4 and 5," op. cit. Ref. 9, pp. 200–205.
27. A. R. Miller, "Polyurethane Diaphragms in High Power Pulsers", Maxwell Laboratories Note MLR-724 (1978) (unpublished).
28. I. M. Vitkovitsky, "Multi-Channel Switching for High Power Discharges," NRL Memorandum Report 1831 (1967).
29. R. A. Miller, Private communication (1985).

7
Opening Switches Using Arcs

INTRODUCTION

The preceeding chapters have all dealt with *closing* switches which have a capability to handle very high power pulses in single switching operation. In this chapter, circuit breakers, used for *opening* a circuit, i.e., interrupting a current flow, usually with the intent to shift it from one circuit branch to another, are discussed. In electrical power transmission applications, circuit breakers are used to interrupt high fault currents before damage occurs in connected equipment. The breaker must be automatically resettable to protect against the next power surge. In pulse power applications, the initial current flow through the breaker is usually used to charge an inductor and then to commutate the current to the load. While in power transmission applications, the fault clearing time equals that of a few 50 or 60 Hz periods, or perhaps a fraction of a period (required in more modern power networks), pulse power applications require faster commutation time. Fast commutation requires the high opening switch voltage necessary to transfer the current to the load. That is, the circuit breaker functions as a power amplification device.

The real need for high power opening switches arose when it was realized that capacitor energy storage banks, while capable of delivering high power, are too large and too expensive to provide also large amounts of energy to the load[1].

Multimegajoule pulses of electromagnetic energy at high power levels are required in a variety of experimental and development programs. For example, in the intertial fusion program, both the laser output and charged particle (electron or ion) beam output must provide power into the fusion target at the tens of terawatts level[2]. Studies such as those involving the simulation of effects of natural lightning on aircraft and other vehicles also require large energy pulsers with output in the subterawatt range[3]. Inductive storage for such applications is very attractive because of its inherent compactness associated with energy storage in the form of magnetic fields. Additionally, inductive storage can be sufficiently versatile to provide high peak power pulse trains for those applications in which a sequence of events separated by only tens or hundreds of microseconds must take place. The appropriate circuits and candidate switches for repetitive pulsing are described in Chapter 9.

Storage of energy in magnetic fields is practical at a level of the order of 10^7 J/m^3. The current required to produce the magnetic field can be derived from sources such as capacitor banks and chemical storage (batteries and explosives), and from inertial storage such as alternators and homopolar generators. With the exception of the capacitor banks, the energy density of these current sources is comparable to that of magnetic fields. Some of these, e.g., explosive generators[4], are capable of producing high current output with very short risetime. Although explosive generators are very compact and are suitable for very large pulse energy, they cannot be used in many environments (and often destroy the experiment) where periodic pulsing of the system is required. Nevertheless, development is progressing toward a significant goal of pulse power research, namely, the attainment of energy per pulse of 1–10 MJ at power levels of 10–100 TW[5]. A more practical, nondestructible, high current pulser is the PULSAR[6] employing explosively generated plasma for compressing magnetic fields.

Rotating machinery, storing energy in intertial form, has been traditionally used as high current sources with either periodic or single pulse output. Electric power test facilities[7] for 60 Hz output and homopolar generators for single pulse output[8, 9] can, in addition, be used with current transformers to adjust output current requirements[10]. More recently, development of hybrid concepts, combining the features of alternators and homopolar generators, has been started[11]. Conversion of the mechanical to magnetic energy, generally, occurs in time several orders of magnitude longer than the required duration of the output pulse.

It is in the context of large energy requirements that the development of very fast opening switches has occurred. This chapter discusses switches and circuit breakers for electric power applications only to indicate typical performance parameters for comparison with the requirements imposed by pulse power applications. To some extent, the development of fast opening switches rests on the technology of the electric power industry. However, the fast breakers, as long as they handle high power, depend almost exclusively on the use of explosives to obtain appropriate switching parameters. One exception to the use of explosives is the ultra-fast circuit breaker, which uses magnetic pressure to cut a heavy conductor[12]. It has achieved results which qualify it for high power applications where the switch voltage does not exceed 100 kV. It is described at the end of this chapter. The development of switches using explosives for interruption of current has occurred mainly in the last 10 years, resulting in the demonstration of very successful devices for interrupting currents up to megamperes, and, if necessary, recovery voltages in the submegavolt range. The price paid for such high speed is that, unlike the conventional circuit breakers which are reusable, the EOCB's are single-shot devices and in some cases use very large amounts of explosives[5]. Because there has been little discussion of explosively operated circuit breakers (EOCB's) while power transmission circuit breakers are extensively covered in the accessible literature.[13,14,15], this chapter focuses on EOCB's.

ELECTRIC POWER CIRCUIT BREAKERS

The main function of the circuit breakers used in the transmission and distribution of electric power is to protect systems against short-circuit current surges. The standard electric power service voltages in the United States are 7.2, 13.8 and 34.5 kV in the distribution networks and range from 69kV to 1200kV in the transmission grids. In major 138 kV stations, fault current levels are in the 100 kA range[16]. Basically, circuit breakers consist of a pair of contacts, one stationary and one moveable. A mechanism operates the moveable contact to close or open the circuit. When contacts are in the closed position, the circuit is completed and energized. In service, it operates in this state, carrying the current continuously. When the surge arrives, the mechanism, sensing the change in conditions, moves the contacts interrupting the circuit. This results in the drawing of an arc between the separating electrodes.

An *arc* is a high temperature plasma with high conductivity. Normally, the power dissipated in the arc from the circuit prevents the resistance from increasing to a value that effectively stops the flow of current. Thus some means of quenching the arc are employed to extinguish it sufficiently rapidly to prevent damage to the network downstream. In alternating current circuits, a natural current zero period is provided which facilitates circuit interruption if arc cooling is sufficient to provide a rate of rise of the dielectric recovery for the contact gap which exceeds the rate of rise of the circuit recovery voltage. In direct current circuits, an artificial current zero must be introduced, or the opening switch must provide sufficient arc voltage for an extended time to oppose the current from the source until interruption is achieved. The later case can involve substantial energy dissipation within the opening switch. Historically, arcs in air, oil, water or ablation vapor from plastics have been used by the electric power industry. Quenching of the arc between closely spaced ceramic walls in air, employing magnetic pressure, is the common technique to attain high arc voltage for dc switches. Sulfurhexafluoride (SF_6) and vacuum ambient with carefully degassed electrodes have been used more recently, mainly because of faster dielectric recovery.

Opening switches in the power industry serve a variety of specialized functions. The four main classes are listed below (in the specialized terminology employed by the electric power industry):

1. *No load* or *disconnect switches* meet all dielectric requirements of its voltage class. They carry the load currents and fault currents without overheating or mechanical destruction. The interruption ability is typically limited to a current of a few amperes, such as capacitive current on unloaded transmission lines. Any switch must be able to close in on an existing fault.

2. The *load break switch* is able to interrupt the load current which may range typically from 100 A to 10 KA.

3. The *fault current limiter* commutates the fault current into a parallel resistive path before it has grown to its full magnitude. These devices require a fast (1 ms) opening switch. They can be used in ac and dc circuits as well. The fault current limiter represents an early and commercially successful application of explosives for fast opening switches. It originated in the 1950's at the Technical University of Braunschweig in West Germany (W. Koch, *Use of Explosives for Fast Circuit Interruption*, in German, doctoral dissertation [1957]) and has been in worldwide service since that time.

4. *Circuit breakers,* which are the bulwark of defense, are designed to make and break the full range of currents up to fault current magnitude which can exceed 100 kA. Specific requirements for the recovery voltage must be met.

Large facilities exist for testing utility equipment. Rotating generators and synthetic interruption test facilities exist which create the test conditions as prescribed by national and international standards[7].

The nature of the networks protected by circuit breakers is such that the circuit voltage appearing across the arc during the interruption may reappear across the breaker after the interruption. While the amplitude and time of appearance of the voltage transient differs in various systems, the typical network operation may lead to doubling of the voltage associated with transients having frequencies in the 10 kHz range[13]. To prevent the restrike (i.e., voltage breakdown of the gas which has just managed to cool into relatively non-conducting state) of the breaker, the voltage hold-off of the gas in the breaker must increase at a rate sufficient to withstand the reapplication of the voltage in the period associated with the system transients. This rate of increase of the dielectric strength of the quenched arc, called rate of rise of recovery voltage (rrrv) of the circuit breaker, is an important specification of the commercial breakers.

An introductory description of the circuit breakers, their operation and their role in the electric power networks, appears in *Scientific American*[17]. A rather comprehensive technical review of the high voltage switchgear, with some 170 references, is provided by C. J. O. Garrard[18], who discusses the principles of design of circuit breakers, isolators, switches, surge diverters, fuses, and associated insulating materials. This reference illustrates the variety of circuit breakers and operating parameters, many of which continue to be used to the present day. Some of the more important types are the old bulk-oil breaker, for 7.5 GVA at 175 kV and the low oil content breakers with up to 20 breaks per pole for 700 kV service. An example of the multi-gap oil breaker is shown in Fig. 7-1, reproduced from *Scientific*

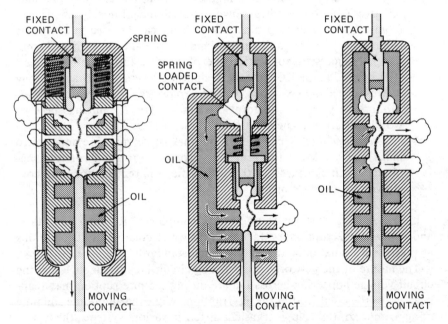

Fig. 7-1. Three types of circuit breakers using decomposition of oil in contact with the arc to cool the arc. The first design (at left) uses blow-out parts for preventing excessive pressure build-up. The second design uses two breaking contacts in sequence, employing the gases generated in the first arc to cool the second arc. In the last case, the moving contact opens the pressure-relieving parts. Reproduced from Werner Reider, "Circuit Breakers," *Scientific American,* p. 83 (Jan. 1971). Copyright © 1971 by Scientific American, Inc. All rights reserved.

American. Other types of breakers, using sulphur hexafluoride (SF_6) gas and air blast as means of arc quenching, and of providing the insulation between contacts, are also reviewed by Garrard[18].

The more recent research issues and circuit breaker development programs are covered, for example, in the Electric Power Research Institute's *Symposium Proceedings on New Concepts in Fault Current Limiters and Power Circuit Breakers*[15]. Some of these issues are concerned with higher current operation. For example J. R. Rostron and H. E. Spindle discuss 120 kA breakers[19]. Other issues deal with faster interrupting times and higher recovery rates. The latter, discussed by A. S. Gilmour[15] and C. W. Kimblin et al.[20], employs vacuum arc breaking. Such vacuum circuit breakers have been developed into 25 kA, 10 kV devices which approach the specifications associated with the applications in high power pulse power technology[20,21]. The lower current vacuum breakers are already mentioned in Ref. 18, suggesting that their operation can be extended by staging of series gaps for voltages up to 500 kV.

Finally, perhaps the most relevant circuit breaker technology that is being utilized in the pulse power applications is that using explosives to cut conducting metal with a resulting fast separation of contacts[22], employed in European distribution networks. Other types and derivatives of such breakers have also been studied for power transmission networks as reported in patents such as that of Kozorezov et al., and that of Dethlefsen listed with other explosively operated circuit breakers (EOCB) in Ref. 23.

An early use of the commercial circuit breakers in pulse power applications, where power amplification was achieved by current interruption of inductive storage coils, is described by J. Salge et al.[24]

EXPLOSIVELY DRIVEN OPENING SWITCHES

Switch Types

Because commercially available circuit breakers and their modifications discussed in the preceeding section are inadequate for many high power, pulsed power applications, high speed opening switches were developed specifically to open in a range of a few to a few tens of microseconds. The significant feature of these switches is the "indefinitely long" (i.e., milliseconds to seconds) conduction time of currents ranging from tens of kiloamperes to fractions of megamperes, as well as high hold-off electric fields (~ 10 to 20 kV/cm) after very short (tens of microseconds) recovery periods. (A special subset of such switches was developed for applications with explosive generators using magnetic compression to generate very high power output[25]. The switches of this subset use large amounts of explosives to open megampere circuits in fractions of a microsecond[26]). While the origin of these designs has been encouraged by the explosively actuated commercial fault current limiters used in Europe[22], the spurt in their development is the result of the effort led by Larinov[27] and as a result of a series of patents[23] granted in the 1970's. This technology was extended to higher voltages[28] and very short opening times, 10 μs (Ref. 29) and 4 μs (Ref. 30) respectively, and applied in the development of high power inductive storage technology.

Two distinct types of switches have evolved from this research. One type depends on the use of an explosive to insert an arc-quenching material, such as paraffin, into the gap under the same high pressure which initially causes the rupturing of the conductor. This type of switch, called here a *mass flow switch,* uses the explosive more economically than the second type and is fully contained in the switch packages (i.e., with no release of switch fragments into the surroundings)[31]. The second type of switch, described in References 29, 30, and 32, ruptures the metal with the explosive in direct contact with the metal. The very high pressure explosive products surrounding the separating electrodes provide the high electric field hold-off and the cooling of the arc to generate rapidly rising high arc resistance. This type of EOCB is called a *pressure-quenched switch.*

The mass flow switch is discussed here more thoroughly than the pressure-quenched switch since many of its characteristics and its role in a circuit are similar. The large explosive mass switch for a submicrosecond opening time, not being applicable to the laboratory environment, is discussed only through the relevant references.

Mass Flow EOCB

As discussed above, the basic concept underlying the mechanical interruption of current is the formation of an arc across the separating electrodes and subsequent cooling of the arc by a contact with cold gases, liquids, or solids in its immediate vicinity, or by temporary reduction in arc current by use of external circuits. Alternately, arc resistance can be used to limit currents to prescribed values. Cooling of the arc leads to increased arc resistance and its eventual extinction. In addition, the separating electrodes and structural elements of the switch, as well as gases and debris generated in the process of disruption, must often withstand the inductive voltage generated across the switch. This is an important requirement which is not easy to meet since, in general, the peak voltage appears during the time the electrodes and the pressure transfer (pusher) medium are still in motion. These are, therefore, two aspects of opening switch design, mechanical and electrical, that must be taken into consideration together.

To study the most important aspects of the explosively actuated opening switch, a basic multi-gap switch module of the mass flow type was investigated by R. Ford and this author[28] to establish switch design data. The switch module is shown in Fig. 7-2. The module is 40 cm long; however its active gap and electrode length is 25 cm.

Mechanical Design. Figure 7-3 illustrates the details of the exploding switch and its functioning. The switch is a cylindrical device, shown before and after the interruption of current. This switch is constructed around the current-carrying aluminum cylinder. The cylinder is filled with inert material with a 50 grain/foot PETN detonating cord in the center. The explosive is initiated by an exploding bridge wire (EBW) initiator at one end of the switch module. The first function of the inert pusher material (paraffin) is to transmit the pressure required to burst the cylinder in those places where no externally mounted steel rings are emplaced. Every other ring is rounded as indicated in Fig. 7-2. This facilitates bending of the aluminum cylinder section so that it lies flat against the ring surface maximizing the separation between rings. The second function of the inert medium is to cool the arc. One of the best choices to perform both of these functions is paraffin[33] as discussed later in connection with electrical performance. For pulsed operations, associated with inductive storage systems, water can also be used, which suggests a method for simplifying switch replacement. Water provides a shorter delay time, about 50 rather than 70 μs, as when paraffin is used, in agreement with the results given in Ref. 33.

Figure 7-4 shows an early exploding switch assembly diagrammed in Fig. 7-2. This assembly, 40 cm long, stays intact even when used for dissipation of energy up to 80 kJ, when the entire inductor energy is dumped into the switch, as is the case with open circuit tests. Figure 7-5 is one of the series of steel rings with the aluminum cylinder section uniformly folded over it by the explosive pressure. To obtain proper metal-forming[31], the aluminum cylinder must be scored axially on the outside surfaces at 20° (or smaller) intervals; gap length and aluminum thickness must also be properly selected. Faulty metal-forming results in curling of the folded metal, effectively reducing the gap. A fast-framing camera has shown that rupturing occurs simultaneously in all rings. The aluminum alloy 6064T6 cylinder (6.5 cm in diameter) with 0.89 mm wall thickness ruptures, producing 16 gaps (25 cm in total length) within 20 μs. This is consistent with the electrical measurements reported below. Thicker walls (1.65 mm) have also been tested to provide data on switch design for higher current and/or longer pulse time operation. Burst simultaneity of gaps and electrical performance remain the same as for 0.89 mm wall cylinders. The 1.65 mm wall thickness can accommodate 100 kA current flow without substantial heating for sig-

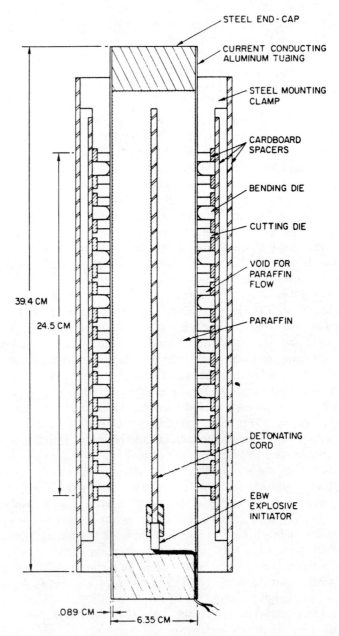

Fig. 7-2. Early design of an explosively actuated opening switch assembly. Reproduced from Ref. 28.

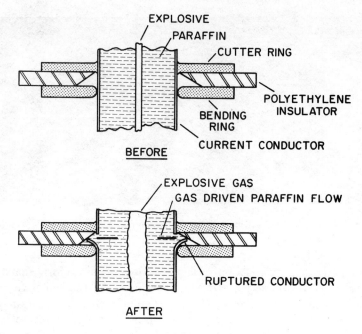

Fig. 7-3. Schematic diagram of explosively actuated switch before and after the switching. Reproduced from Ref. 28.

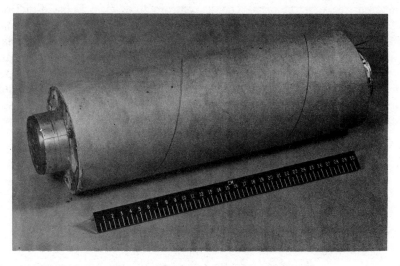

Fig. 7-4. EOCB module fully assembled. Reproduced from Ref. 28.

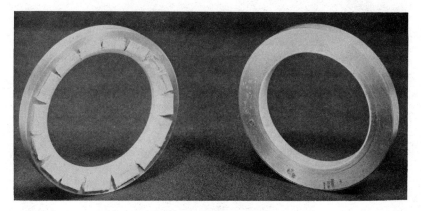

Fig. 7-5. Results of explosive metal forming (at left) onto a ring shown at right. Reproduced from Ref. 28.

nificant periods. Larger, as well as smaller, amounts of explosive have been also tested to determine the effect of the amount of the explosive on the switch opening. To maintain good rupturing uniformity, the location of the explosive initiator at one end of the switch module plays a key role. Premature venting of the explosive gas pressure also leads to slow or incomplete rupturing.

The time delay between the initiation signal and the cylinder rupturing is given in Table 7-1. This time decreases with explosive charge weight. However, when charge weight exceeds 8 g the steel rings begin to yield and tempered steel must be used. The delay, from 40 to 75 μs as listed in Table 7-1, is the time required for the explosive to detonate, compress the paraffin, fill available voids, and rupture the aluminum tube at the appropriate ring dies. After the tube is ruptured, it separates along axial scribe lines and folds around the bending dies as seen in Fig. 7-5, at a rate depending upon paraffin flow. For design purposes, the paraffin velocity may be calculated approximately from test parameters listed in Table 7-2 and from the succeeding equations.

Table 7-1. Time Delay Between Initiation and Switch Opening.

Wall Thickness (mm)	Weight of Explosive (gm)	Time to Rupture (μs)
0.89	3.24	70 ± 5
0.89	5.67	50 ± 5
0.89	8.10	40 ± 5
1.65	5.67	140*

*Minimum time to rupture has not been determined for switches using 1.65mm wall thickness.

Table 7-2. Development Switch Module.

Explosive charge, M	5.67 gm
Explosive (PETN) energy, W	4200 J/gm
Paraffin mass, m	0.82 kg
Initial explosive volume, V_1	9.5 cm³
Final explosive gas volume, V_2	390 cm³

Gas efficiency[34, 35], η, defined as a ratio of gas energy at venting, E_F, to explosive potential energy, E_I, gives (using specific heat ratio, $\gamma = 1.2$):

$$\eta = \frac{E_F}{E_I} = \left(\frac{V_i}{V_f} \right)^{\gamma - 1} = 50\% \qquad (7\text{-}1)$$

The efficiency of transferring the internal energy of the gas to the kinetic motion of a pusher is estimated to be about 50% on the basis of results obtained in a test assembly[35] using similar explosive weight, gas volume, and pusher weight. Thus, the overall efficiency, η_{TOT} of 25%, suggests that the pusher velocity, v, given by

$$v = \sqrt{(2\eta_{TOT} W)/m} \qquad (7\text{-}2)$$

is 0.12 mm/μs for the parameters in Table 7-2. The estimated velocity agrees with measured values discussed below.

Electrical Properties. The mechanical design of the switch leads easily to a variety of switching functions. Stacking of many sections can be used to maintain a large voltage across the switch; the conductor thickness or parallel use of switches is effective in carrying a large current for an indefinite time before interruption; the opening time can be decreased by loading the switch with an increased amount of the explosive. Triggering is quite precise, i.e., triggering jitter is substantially shorter than the opening time. The EBW detonators can be initiated with submicrosecond accuracy, as already pointed out in Chapter 5. The high detonation velocity (7000 m/s) provides rapid pressure build-up resulting in a high degree of simultaneity of gap rupturing. These properties combine to give a low dissipation energy switch for high voltage, high current operation with command trigger capability. Arc voltages of 0.5 to 1.0 kV/cm are typical for this type of current interruption.

Breakdown Properties. The voltage that can be maintained across any given switch depends on the choice of the dielectric and the design of the switch and on the switch's arc properties. The pusher medium, such as the paraffin, serves as an insulator to prevent discharges between the cutting and bending rings. In the process of rupturing the current-carrying cylinder, sharp jagged edges of the order of a fraction of a millimeter are formed (as seen in Fig. 7–5). The ring thickness and its detailed surface-shaping as well as the gap length between rings are chosen to allow the aluminum to follow the contour of the ring, optimizing the gap length. Thus, the electric field averaged over the length of the switch, determined by the ring and gap lengths, can be high when the gaps are filled by the paraffin.

There are other possible breakdown paths that can be initiated by the voltage arising during current interruption. One of these paths can initiate on the inner surface of the expanding paraffin cylinder and puncture to the bending or cutting rings. Another breakdown path can develop on the outside surfaces of the switch. High voltage tests[28] with switch element stacks ranging in length from 4 to 25 cm have shown that the weakest insulation is between the outside edges of the cutting rings. It is on the surface of the paraffin which is moving against the confining cylindrical (cardboard) spacers, as shown in Fig. 7–2.

The high voltage tests separating the mechanical functioning of the switch (the explosive disruption of the conducting cylinder and paraffin flow) from the electrical function (the interruption of current and subsequent buildup of the voltage across the switch) via an external test circuit have shown that up to 100 kV in 10 μs can be applied across the switch module, as illustrated in Fig. 7–6. Letter E in the circuit in Fig. 7–6 denotes the EOCB element undergoing the tests. By the employment of this external circuit, the potential could be applied at different times ranging from periods before switch rupture initiation to periods several minutes after explosion. In these tests, no current was carried by the switch so that the separation of electrodes occured without formation of an arc. As can be seen in Fig. 7–6, the switch can hold the voltage indefinitely or for short periods (about 100 μs), depending on the time of application of voltage relative to conductor rupturing and on the average electric stress associated with a given switch length. Data points associated with short hold-off periods are denoted by asterisks. The applied voltage waveform is also shown in Fig. 7–6.

The synthetic testing method, depicted in Fig. 7–6, is a convenient tool in the design of EOCB's. Such testing points out the causes of poor holdoff capabilities and evaluates the opening time without the use of high energy storage pulsers. Additionally, full scale testing in the initial stages of development tends to obscure the post-mortem analysis. The results of tests show that a 25 cm long switch design, using 16 ring-gap sections, can be stressed at least to 3.3 kV/cm. Since the degree of gap-opening simultaneity may affect the breakdown stress level, shorter switches have also been tested. These tests indicate that up to 10 kV/cm can be achieved for a two-section switch with this design if a breakdown can be tolerated after some 20 μs.

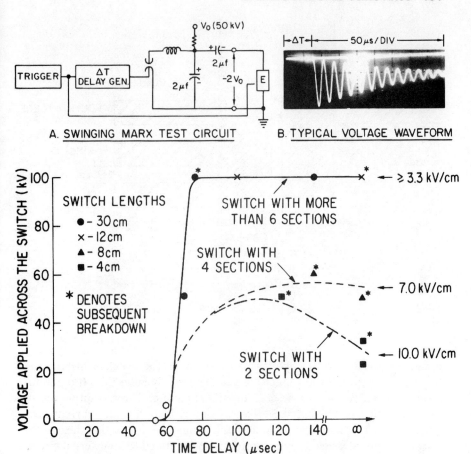

Fig. 7-6. EOCB voltage hold-off test results. Reproduced from Ref. 28.

Once the breakdown path was determined to be between the outer rims of the rings, it was possible to redesign the insulation and electric field distribution as shown in Fig. 7-7. This configuration has been tested under the arc conditions described below.

In addition, the switch design shown in Fig. 7-3 has been tested using voltage pulses with decay time exceeding 100 ms. These tests, confirming the results for a 16-section switch, are given in Fig. 7-6, indicating that this data is valid in applications related to power transmission, for which extended dielectric integrity is required after opening.

Arc Properties.   The characteristics of the EOCB while interrupting the current can be divided into several categories. The three most important

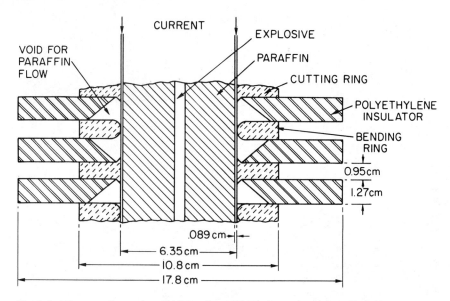

Fig. 7-7. Diagram of a section of high voltage EOCB. Reproduced from Ref. 28.

ones are associated with its performance where (1) the switch is interrupting the current with a period much longer than the opening time (i.e., simulating the interruption of dc current); (2) the switch is interrupting an alternating current, so that the effect of arc extinction under current zero conditions can be determined; and (3) the switch operates so that the current during the initial phase of interruption is commutated to an external circuit branch with an exploding wire (fuse) for generating higher voltage. In applications in which more than one switch module is required, small jitter in the delay between the initiation of the explosion and the current interruption is also an important characteristic.

Fig. 7-8 shows the typical behavior of the rise in arc resistance during the interruption of current in the 25 cm long exploding switch using different amounts of explosive. The change in arc resistance arises from the increase in the gap formed during the rupturing of the current-carrying cylinder. In these experiments, the current in the switch was 20 kA. The rate of growth of the gap length, directly proportional to the rate of growth of the arc length, is calculated using a simple model shown in the insert in Fig. 7-8. This model represents the section of the aluminum cylinder as a straight flap hinged at one of the bending dies, moving with a velocity determined from Eq. 7-2 for a given amount of explosive. The calculated velocity agrees with the measured values obtained with the use of a framing camera to

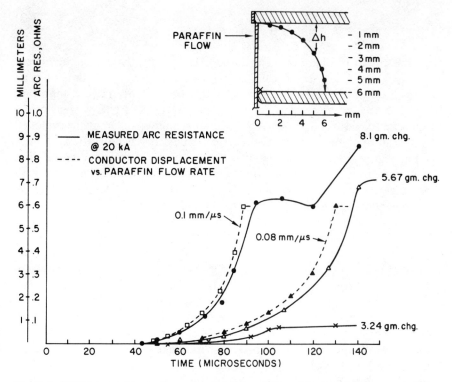

Fig. 7–8. EOCB arc resistance for various explosive charges. The insert at the upper right shows the folding of the metal, cut by the high pressure. The position Δh is also indicated on the ordinate, correlating the folding position and the arc resistance. Reproduced from Ref. 28.

observe the motion of the aluminum. This good agreement indicates that the perturbation of the flow by the friction with the rings is not significant.

The almost exact correlation of the two sets of curves in Fig. 7–8, the gap length, and the resistance of the arc for a range of charges varying by a factor of 2.5 indicate that during this time the developed voltage and its rate of rise depend only upon the gap length and the rate of gap opening respectively. This is similar to behavior in a gas-cooled arc with gap opening rates of the same order of magnitude[36].

The current-voltage characteristics of the arc exhibit two types of behavior, as can be seen from the oscilloscope traces in Figs 7–9 and 7–10. Initially, the voltage is rising rapidly. The current in the switch, at the end of the voltage rise time is essentially unchanged as can be expected from the equation

$$V(t) = RI = L \frac{dI}{dt} \qquad (7\text{–}3)$$

for the circuit in Fig. 7-10. For an 80 μH inductor, the change in current is only 2-3 kA, i.e., some 10% of the initial current, at the time of peak arc voltage of about 10 kV. This relatively little change is very important in inductive storage systems, in which the efficiency of energy transfer depends on the ratio $(I/I_0)^2$, where I_0 is the current interrupted by the exploded switch and I is the commutated current. The arc behavior after the peak voltage corresponds to the instant when the paraffin is expanding between the rings, after the metal has folded. At this point in time, the arc length becomes constant and its behavior follows Mayr's model[13] described by the balance between the input power, IV, and losses, W_0:

$$\frac{dQ}{dt} = IV - W_0 \qquad (7\text{–}4)$$

The arc behavior follows the model for non-evaporating plates[13] with separation equal to that between the die rings shown in Fig. 7-2, rather than the separation equal to the total of arc length sections.

The arc resistance rises to an approximately constant value in a time which depends on the amount of the explosive charge as shown in Fig. 7-8. As the current being switched increases, the voltage developed across the arc (the arc voltage) does not change for a given amount of the explosive. This result is summarized in Figure 7-10 for the basic 25 cm module as well as for other configurations. For example, data for two modules connected in series shows that the voltage output is doubled relative to the single module.

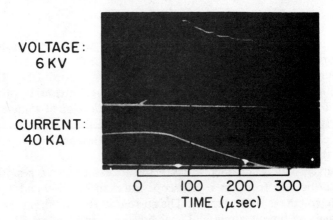

Fig. 7-9. Arc current and voltage waveforms resulting from interruption of current in 80 μH inductor. Reproduced from Ref. 28.

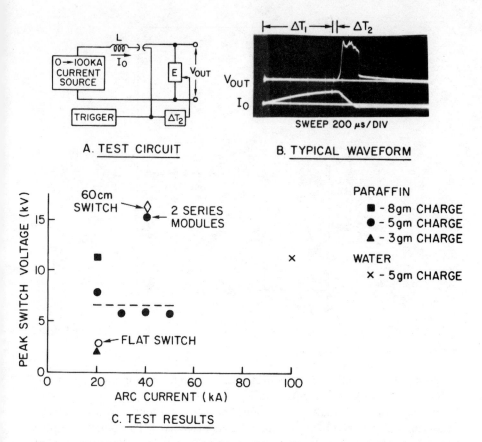

Fig. 7-10. EOCB arc voltage for several explosive charge values and for different pressure transfer media. Reproduced from Ref. 28.

Increasing the module length to 60 cm also results in an approximate doubling the output voltage. Such long modules, furthermore, retain the same opening time so that rate of rise of voltage is also doubled.

Use of the switch at higher currents has shown marked deviation from the scaling based on the results obtained at currents of less than 100 kA. Figure 7–11 shows arc resistance recovery as a function of the time between full current interruption and application of high voltage[28]. The arc resistance can be determined by measuring the current re-established in the switch due to the reapplied voltage. The rate of recovery depends on the current level being switched. At high currents, the recovery rate is too slow for efficient storage system designs. As the data of Fig. 7–11 and the results shown earlier indicate, the recovery rate is very fast for small currents. This

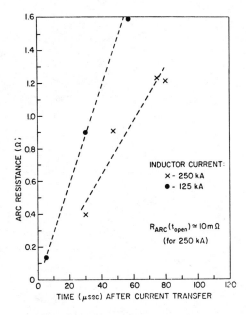

Fig. 7-11. Arc resistance during the interruption of large currents. Reproduced from Ref. 28.

suggests that to achieve fast recovery rates, switch opening must occur under conditions of no current.

To combine both features, i.e., fast recovery and high current interruption capability, a sectioned switch has been developed[28]. This switch is shown in Fig. 7-12. The first 16 gaps are opened 40 μs before the second set of 27 gaps is opened. Two separate initiator caps controlled by a delay generator are used for this purpose. The gaps that open first limit the current to a low value (or, alternately, permit the current to be commutated to a second stage fuse) so that gaps opening later open at low current level. The resulting electrode burn is intense for gaps 1-17 and almost nonexistent in the remainder of the gaps. The two-section switch shown in Fig. 7-12, interrupting currents up to 400 kA after 150 μs conduction, has an arc resistance higher than 100 Ω at $T_i = 50$ μs. While the 27 gaps do not open simultaneously, the opening time is sufficiently short to generate a large inductive voltage. This switch design has been also adapted for high voltage operation, discussed later in this section.

Because the jitter in the time between the explosive initiation and the mechanical rupturing of the conducting cylinder is much shorter (≤ 10 μs) in comparison with voltage buildup time (about 40 μs in Fig. 7-9), the switch can be modularized into various configurations. When, for example, a se-

GROUND

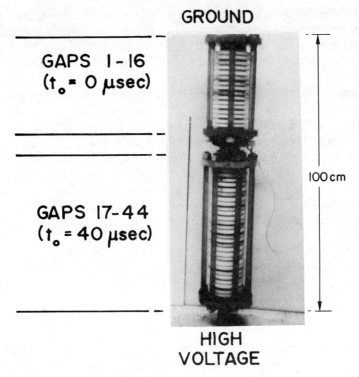

GAPS 1-16
(t_o = O μsec)

GAPS 17-44
(t_o = 40 μsec)

IOO cm

HIGH
VOLTAGE

Fig. 7-12. Two section EOCB with programmed initiation. This switch uses dielectric tie-rods to prevent premature release of pressure. Reproduced from Ref. 28.

ries two-module switch is used, the arc voltage is very nearly twice that of the single module switch (Fig. 7-10). The power level in this test exceeds 600 MW, achieved in 40 μs without use of commutation or of external circuits to generate current zero.

Restrike Properties. Application of the exploding switch in power transmission networks or in production of high power pulses based on inductive storage systems requires voltages as high as 1 MV across the switch. Therefore, in addition to proper insulation as discussed earlier, the switch voltage recovery rate must be consistent with system requirements. The recovery of the EOCB achieved immediately after the rupturing of the cylinder, during the use of a 60 kV, 20 kHz pulser generating peak current of 20 kA through the switch before interruption, was more than 1 KV/μs some 40 μs after current interruption was initiated. This value is intermediate between the values attainable by commercial SF_6 breakers and vacuum interrupters[37] and agrees with that of the exploding switch described in Ref. 9. The voltage recovery tests were performed during the phase of the switch operation when the separation of the aluminum is not yet completed, i.e., within less than 40 to 50 μs after the rupturing of the conductor.

Current Commutation. The high voltage hold-off of the exploding switch module (demonstrated in the tests under conditions where no current was flowing) and good rrrv values observed after current zero suggest that high current, high voltage operation can be obtained by commutating the current from the switch to external circuits. This, indeed, has been demonstrated, with the use in the external circuits of fast-acting fuses (copper wires or aluminum foils, described in Chapter 8) with resistance substantially higher than the 20 $\mu\Omega$ resistance of the exploding switch[28]. The current commutated through the fuse was interrupted in a time shorter than the exploding switch's opening time and with higher final resistance, allowing much higher output voltages to be generated across the switch-fuse combination than those obtained using an exploding switch alone. The details of such two stage switching, based on sequential opening of parallel switches, are elaborated in Chapter 8.

Multiple Module Testing. The exploding switch has been tested in configurations of two modules connected in series and in parallel[28]. These tests were performed to provide data on jitter in the opening time of each module and the effect of jitter on the electrical performance of the switch arrays. Two tests, shown on Fig. 7–13, were performed. The circuit diagram in Fig. 7–13A shows how two modules were operated to double the output voltage to a level of 60 kV. Initially, the current interrupted by the exploding switch was commutated to the exploding Cu wire denoted as the variable resistor. The wire parameters were selected to generate a 30 kV open circuit output. The combined 60 kV output voltage is shown in the top trace. In addition, current in each fuse was included for determining the jitter time. The resultant jitter time of about 10 μs is typical for 25 cm modules. With better control of pressure release, jitter time could be reduced to 1-2 μs.

To provide a pulse train output[28] two switch modules can be connected in parallel with sequential interruption of current provided by appropriately delayed triggers, as shown in Figure 7–13B. This technique is discussed in Chapters 8 and 9. In the case of the EOCB's, the minimum separation between pulses in the train is controlled by the combined effect of switch opening time (of less than 20 μs) and jitter time (10 μs).

The simple design of switch modules allows a relatively large number of them to be used in applications for which pulsing rates are sufficiently low to allow for switch replacement time.

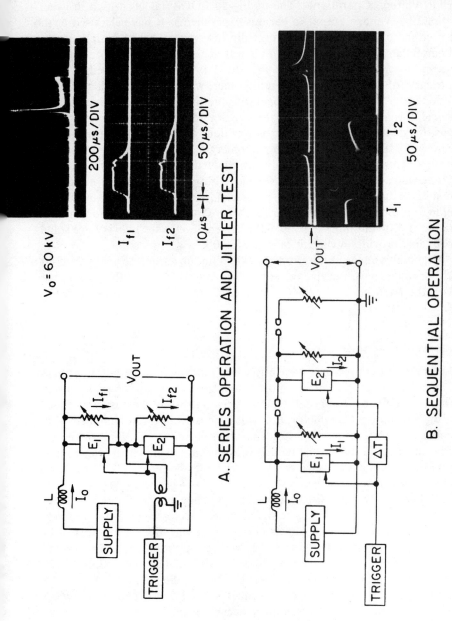

Fig. 7-13. Use of two EOCB modules (A) to double the output voltage from an inductive store and (B) sequential operation of parallel modules to generate closely-spaced burst of pulses.

High Voltage Operation.   The multi-gap EOCB has been used in generating high voltage output in various experiments. It has been used at the lower end of the high voltage spectrum to generate 200 kV output from a homopolar-energized inductor. This was achieved by use of a fuse for the second stage opening switch to provide 35 kA with 100 μs risetime into a primary of current step-up transformer with a 10^{-3} Ω load in the secondary[10]. To increase the operating voltage toward the submegavolt range, it is necessary to insulate the switch externally by using high pressure gas, by submerging it in oil, or, as was done in the Naval Research Laboratory TRIDENT program[31], by submerging in demineralized water. A switch shown in Fig. 7-12 has been adapted for operating under water. Using water as insulation, it is possible to obtain 600 kV output pulse when a suitable fuse is employed for the second stage switch. This switch, operated in the mode discussed earlier, where the two series sections are initiated at different times to control the current level in the high voltage standoff gaps, provides a current output of 400 kA.

The high voltage studies, up to 200 kV in air and up to more then 600 kV in water, have not uncovered any inherent difficulties that would prevent the extention of the operating voltages into the multi-megavolt regime for restrike fields of about 10 kV/cm.

Pressure-Quenched EOCB

In an effort to improve storage efficiency and simplify operation of pulsed inductive storage systems, a simple switch based on modified explosive cutting of conductors has been developed. In contrast to the mass-flow EOCB using a cylindrical conductor, this switch consists of a flat aluminum sheet corrugated to form as many folds, such as shown in Fig. 7-14, as needed for voltage hold-off. It is intended for applications in which long conduction time may be necessary and voltages typically in the 100 kV range are required. The opening switch shown in Fig. 7-14 is a single gap device and utilizes the very high dielectric strength of high pressure gases generated during the explosive cutting of the metal conductor[25, 26], obviating a need for paraffin. The success of this device has led to the development of derivative multi-gap designs for higher voltage applications[30]. The principle of operation of this switch is based on studies such as that of Korolkov et al.[39] showing that the dielectric strength behind the detonation wave (i.e., at times > 1 μs after the front passage) is 1.0–1.5 MV/cm until (some 3–5 μs later) a significant pressure drop occurs. Such high pressures have been used to shut off completely 35 kA of current in 2 μs without commutating the current into another path. This was achieved by use of an explosive in contact with a thin (0.05 mm) Cu strip, generating an inductive pulse of 8

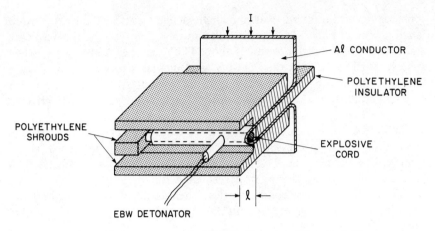

Fig. 7-14. Pressure-quenched EOCB. Reproduced from Ref. 29.

kV. These results suggest that while an interval longer than the 2 μs associated with the experiments of Voytenko et al.[32] is required for effectively separating thicker (0.5–1.0 mm) aluminum sheets (needed to extend the current carrying time to a range of 0.1 to 1 ms), it is possible to maintain the pressure of the gas derived from the explosion at a sufficiently high level for a period necessary to cut thicker sheets and to provide the conditions required for relatively high recovery voltage. In contrast to results reported in Ref. 32, where the metal appears to be vaporized by the explosive, the recovery of a switch using thicker conductors is determined in part by the time required for the metal (cut rather than vaporized by the explosive charge) to move a sufficiently large distance to hold-off the voltage.

Figure 7-14 shows schematically a thick conductor switch configuration designed to show the effects of pressure confinement on switch hold-off voltage. It consists of an aluminum conductor 1 mm thick by 15 cm wide wrapped around a 0.6 cm thick polyethylene sheet and clamped between two thicker polyethylene sheets. The latter polyethylene sheets serve as a shroud (overhang) of length ℓ to partially confine the high pressure explosive products. A single channel arc formed by the interruption of the current occurs as a result of metal separation at a point farthest away from the detonator which sets off the explosive cord. The arc reaches 2 kV in 10 μs and remains at that level for about 50 μs, for $\ell = 1.5$ cm. Photographs indicate that the arc path follows the surface of the expanding high pressure gas volume. For $\ell = 0$ cm, explosive pressure drops more rapidly and the arc voltage collapses after 20 μs. In addition to the arc voltage duration, switch hold-off voltage, measured by applying an external pulse at time t_p

after explosive initiation, was shown to depend critically on the shroud length ℓ seen in Fig. 7–14. Figure 7–14 shows at $\ell = 0$ the hold-off voltage associated with the high pressure is about 48 kV/cm (i.e., 28 kV/gap). When ℓ is extended to 1.5 cm, the hold-off level increases to 60 kV/cm. Figure 7–15 also shows the time dependence of the breakdown voltage for $\ell = 1.5$ cm, indicating that the explosive product insulation remains effective at least for 80 μs.

The hold-off level increases to 100 kV/cm when an end shroud is added to contain explosive pressure produced as the detonation reaches its extremity. The data in Fig. 7–15 also indicate that the opening time is 5–10 μs after explosive initiation, in agreement with the arc voltage rise time mentioned above. These characteristics, when used to derive pressure decay time, indicate good agreement with the expected dynamics of the explosive products.

The recovery voltage of the switch with $\ell = 1.5$ cm (Fig. 7–15) was measured at the 25 kA level (7 kA per cm of the switch width), diverting the current to a fuse. At a time of 35 μs (i.e., at 60 μs after initiation), arc

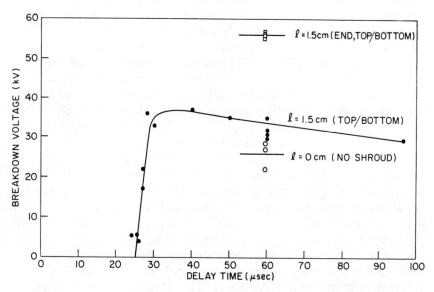

Fig. 7–15. Pressure-quenched EOCB recovery (with no current flow) shown as a function of the delay time relative to the explosive initiation. Data are shown for various degrees of confinement. The lowest recovery voltage occurs with no shrouds, increasing as a result of adding top and bottom shrouds, reaches 60 kV with the addition of end-shroud confinement. Reproduced from Ref. 29.

recovery voltages were the same as the breakdown voltages shown in Fig. 7-15, indicating no degradation of electrical properties due to arc heating.

The switch described here is easy to use in experiments for which no other current interruption methods are practical. Because of its long current-carrying capability, it is suitable for charging inductors using homopolars as current sources. Test of the explosively actuated parallel switching using paraffin indicates that switch jitter depends primarily upon explosive coupling to mechanical components and may be kept as low as ~1 μs as mentioned above, with the current shared by both switches. This feature implies that pressure quenched EOCB can be used in parallel for higher (megampere) current regimes. The amount of the PETN explosive (50 grain/ft) is not large, i.e., 1.6 g for the 3 cm wide switch shown in Fig. 7-14. The switch provides four times higher arc voltage and six times higher recovery voltage (and at an earlier time) than the cylindrical switch configuration using paraffin for cutting the conductor.

Pressure quenched EOCB can be operated without fuse stages in applications where the current is commutated into low resistance loads. P. S. Levi et al.[30] have developed an EOCB consisting of four sections using a 0.25 mm aluminum sheet conductor. The switch is opened with a 175 grain/ft detonator cord, interrupting peak currents of 185 kA after about a 70 μs conduction period. With an opening time of 4 μs obtained by using more explosive per unit length of the switch, a switch voltage of 50 kV across four gaps can be maintained during current transfer to the load. The lower hold-off voltage per gap is due to simpler (and cheaper) construction of the switch, resulting in earlier pressure decay. The behavior of arc resistance during the opening was measured carefully to provide design data for scaling of such switches to higher currents and voltages[30]. Typically, the rate of increase is 117 mΩ/μs, leveling off to 0.5 Ω at the time of peak current. As the current in the switch decreases due to transfer to the load, the explosive products extinguish the arc; and the switch resistance begins another, although slower, rise to virtually open circuit conditions.

Ultra-Fast EOCB

The development of devices driven by chemical explosives to generate megampere currents by magnetic flux compression has been studied extensively[25]. Attempts to use such "explosive generators" as part of a complete pulse power system have led to development of yet another type of EOCB. It was recognized that, because flux compression generators use large amounts of explosives, the practicality of such devices will not diminish if additional amounts of explosives are used to operate switches. Since the opening switching time must be very short for best use of the magnetic flux compression generators, the conducting sheet has been designed to be very low mass (thinly deposited aluminum on a Mylar sheet) compressed by explosive material with the explosive's mass in a kilogram range. References 25 and 26 describe the designs and switch performance. Typical operating parameters listed in Ref. 26 show the opening time to be about 0.5 μs at 3.5 MA with 60 kV across the switch. Modeling of the switch performance[5, 40] indicates a good understanding of the switching operation.

EOCB Commutation

The EOCB's described in this chapter operate very efficiently in a sense that the energy required to accomplish current interruption does not come from the source of the current, but is stored in the chemical explosive. In pulse power systems using inductors, the switching efficiency of commutation of the current from the circuit breaker to the load can be defined as the ratio of the energy dissipated in the switch to the energy stored in the inductor charged via EOCB. Switch dissipation energy is the power dissipated in the switch resistance during the time of commutation. Because the commutation time is a function of arc behavior—and commutation time is required for assessing the performance of the inductive storage systems using EOCB's—the time to transfer the current to the load is now calculated. It is assumed, following Ref. 28 (e.g., NRL Memorandum Report 4168), that the load consists of resistance, R, and inductance, L; the current transferred to the load is driven by the arc voltage, V, developed by the interruption of current i_1. It is given by

$$V(t) = \begin{cases} V_{max}(t/t_e)^2 & t < t_e \\ V_{max} & t > t_e \end{cases} \qquad (7\text{-}5)$$

where t is time measured from the onset of opening of the switch, and t_e is the "opening time." This voltage dependence is valid for mass flow and pressure-quenched EOCB's. For the mass flow EOCB's discussed here, t_e is about 80 μs and V_{max} is 0.5 kV/gap times the number of switch gaps. For a switch with 26 arc gaps, $V_{max} = 13$ kV. As long as the arc voltage is independent of the current flowing through the explosive switch, the resulting load current can be determined without regard for the current or other parameters of the storage inductance, L_0, part of the circuit.

The transfer time, t_T, is given by the circuit equations for two time regimes of Eq. 7-5, because it is not possible to tell a priori whether t_T will exceed t_e.

$$\frac{di_2}{dt} + \frac{R}{L} i_2 = \begin{cases} (V_{max}/L) \ (t/t_e)^2 & 0 < t < t_e \\ (V_{max}/L) & t > t_e \end{cases} \qquad (7\text{-}6)$$

This equation can be solved for the time dependent load current $i_2(t)$, as in Chapter 8 for the constant value $V(t)$. Similarly as in that chapter, commutation of the current from the explosive switch will be achieved if $i_2(t)$ increases to a point where it equals the storage coil current, i.e., the condition for commutation at a time t_T is $i_2(t_T) = i_0$. Here i_0 represents the initial value of the storage coil current. If L_0 is very large, i_0 is essentially a constant. If L_0 is smaller, so that its current is variable, then i_0 is the value of coil current measured at initiation of commutation.

The simplest case for calculation of the load current from Eq. 7-6 occurs when the series switch, at the output of the inductor, closes after the explosive switch arc voltage has reached V_{max}. If the series switch closes at t_1 (measured from the start of explosive switch opening) and $t_1 > t_e$, then the right-hand side of Eq. 7-6 is a constant, and for $t < t_1$

$$i_2 (t) = \frac{V_{max}}{R} \left[1 - e^{-\frac{R(t - t_1)}{L}} \right] \qquad (7\text{-}7)$$

This equation shows that $i_2(t) < V_{max}/R$ so that full commutation of the current can be achieved only if

$$R < \frac{V_{max}}{i_0} \qquad (7\text{-}8)$$

This condition must be satisfied in all cases, since V_{max} is the greatest possible voltage and the fuse current can never exceed V_{max}/R. Using the value of the current at t_T specified above, Eq. 7-7 can be inverted to obtain the dependence of time of current transfer on the circuit parameters. This is further discussed in Chapter 8 and graphically portrayed in Fig. 8-3.

MAGNETICALLY RUPTURED OPENING SWITCHES

Fast opening of switches based on mechanical rupture of current-carrying conductors has also been explored. Two types of interruption are used to obtain opening times of about 10 μs. In one developmental version[38], an oil cavity is pressurized by bursting a submerged coaxial conductor with large magnetic pressure. The pressure pulse is caused by a command current pulse. The oil pressure is sufficient to break the conductor carrying the main current pulse in the conductor. Further development of this version has been stopped upon achievement of interruption at 200 kA. One drawback of this switch is the necessity to use a very high command current pulse (of short duration) of about 500kA.

A different version[12] has been tested using magnetic pressure more directly. The operating force for this type of EOCB is obtained from the magnetic pressure produced by two parallel conductors carrying the main current (in the opposite directions) by introducing a short command current (again, greater than the main current to be interrupted) in such a manner that it opposes the current in one conductor and adds to the current in the second conductor. Tests have been limited to the 40 kA level. The restrike voltage of 40 kV has been obtained using 25–50 bar pressure in N_2, surrounding the arc. At this current level the 8-gap arc voltage at the switch terminals was only 1 kV.

The main disadvantage of magnetically operated EOCB's, in addition to the required high current command pulses, is the difficulty in adapting such switches to high voltage operation because the connections of trigger circuit components operate near the ground level.

REFERENCES

1. S. A. Nasar and H. H. Woodson, "Storage and Transfer of Energy for Pulsed Power Applications", *Proceedings of the Sixth Symposium on Engineering Problems of Fusion Research,* San Diego, CA, IEEE Pub. 75CH1097-5-NPS, pp. 316–321, (1975).
2. C. M. Stickley, "Laser Fusion," *Physics Today* **31**, 50–58 (May 1978).
3. M. A. Uman, "Comparison of Lightning and a Long Laboratory Spark," *Proceedings of the IEEE,* **59**, 457–466 (1971).
4. H. Knoepfel, *Pulsed High Magnetic Fields,* North Holland., Amsterdam and London (1970).
5. While explosive generators are described in many sources, as for example in Ref. 4, the interruption of current to increase the output power from inductors is described in less widely available references. The primary sources dealing with large inductive storage are those of A. I. Pavlovskii, U. A. Vasyukov, and A. S. Rusakov, "Magnetoimplosive Generators for Rapid-Risetime Megampere Pulses," *Sov. Phys. Tech. Phys. Letter* **3**, 320–321 (1977) and B. N. Turman, T. J. Tucker, and P. J. Skogmo, "Magampere Current Experiments with the Intrepid Fast Opening Switch," Sandia Report SAND 82–1531 (1982) as well as Ref. 40 below.
6. M. Cowan, E. C. Cnare, W. K. Tucker, and D. R. Wesenberg, "Multimegajoule Pulsed Power Generation from Reusable Compressed Magnetic Field Device," *Proceedings of International Conference on Energy Storage Compression and Switching,* Plenum, New York, p. 131 (1976).
7. BBC Electric High Power Laboratory, Chalfont, PA, can provide up to 1000 MVA for testing of circuit breakers and other power transmission and control equipment.
8. W. F. Weldon, M. D. Driga, H. H. Woodson, and H. G. Rylander, "The Design, Fabrication and Testing of a Five Megajoule Homopolar Motor-Generator," p. 325, op. cit. Ref. 1.
9. A. E. Robson, P. Turchi, W. H. Lupton, M. Ury, and W. Warnick, *Proceedings of International Conference on Energy Storage, Compression and Switching,* pp. 95–103, Plenum, New York (1976).

10. R. D. Ford, D. Jenkins, W. H. Lupton, and I. M. Vitkovitsky, "Pulsed High-Voltage and High-Current Outputs from Homopolar Energy Storage System," *Rev. Sci. Instrum.* **52**, 694 (1981).

11. W. F. Weldon, W. L. Bird, M. D. Driga, K. M. Tolk, H. G. Rylander, and H. H. Woodson, "Fundamental Limitations and Design Considerations for Compensated Pulsed Alternators," *Proceedings of the Second International Pulsed Power Conference,* Lubbock, TX, IEEE Cat. No. 79CH1505-7, p. 76, (1979).

12. C. Caupers, C. Rioux, and F. Rioux-Damidau, "New Type of Ultrafast Circuit Breaker: Its Principle and Performances," *Rev. Sci. Instrum.* **52**, 118 (1981).

13. T. H. Lee, *Physics and Engineering of High Power Devices,* MIT Press, Cambridge, MA (1975). A very good discussion of moving electrode spark gaps is given in "Moving Spark Gaps" by F. B. A. Frungel, in *High Speed Pulse Technology,* Vol. I, Academic Press, New York (1965).

14. G. Frind, *Current Interruption in High Voltage Network,* K. Rogaller, ed., pp. 67–94, Plenum, New York (1978).

15a. Symposium proceedings, *New Concepts in Fault Current Limiters and Power Circuit Breakers,* EPRI Report EL-276-SR, Palo Alto, CA (1977).

15b. A. S. Gilmour, "Feasibility of a Vacuum Arc Fault Current Limiter," Paper 17, op. cit. Ref. 15a.

16. C. A. Falcone, J. E. Beekler, W. E. Makolites, and J. Grzan, "Current Limiting Device— A Utility's Need," IEEE PES Summer Meeting and EHV/UHV Conference, Vancouver, Paper C 73 470-2 (1973).

17. W. Rieder, "Circuit Breakers," *Scientific American* **224**, 76–84 (January 1971) (same as Ref. 1, Ch. 1).

18. C. J. O. Garrard, "High Voltage Switchgear—A Review of Progress," *Proc. IEE* **113**, 1523–1539 (1966).

19. J. R. Rostron and H. E. Spindle, "High Density SF$_6$ Interrupters for 120kA Development," Paper 6, op. cit. Ref. 15a above.

20. C. W. Kimblin, P. G. Slade, J. G. Gorman, F. A. Holmes, P. R. Eustage, R. E. Voshall, and J. V. R. Heberlein, "Developmental Studies of a Current Limiter Using Vacuum Arc Commutation," Paper 18, op. cit. Ref. 15a above.

21. W. M. Parsons, "Design and Testing of a Prototype Water-cooled Vacuum Interrupter for Use in Superconducting Magnet Protection Circuits," Los Alamos National Laboratory, Report LA-9687-MS (1983).

22. H. Von Paul Brueckner, "Ein Neuartiges Schaltgerat mit Ausserst Kurzen Schaltzeiten," *Electrotechnische Zeitschrift A* **79**, 33–35 (1958).

23. Representative EOCB patents with application in electric power transmission and in pulse power circuits are:
    a. A. I. Pavlovskii, R. Z. Ludaev, and V. A. Vasyukov, USSR Patent 455716 (25 July 1975).
    b. K. I. Kozorezov, V. V. Semchenko, and G. I. Mikhaylu, USSR Patent 475680 (30 June 1975).
    c. A. I. Pavlovskii and V. A. Vasyukov, USSR Patent 548707 (15 May 1979).
    d. R. D. Ford and I. M. Vitkovitsky, US Patent 4,174,471 (13 Nov. 1979).
    e. R. Dethlefsen, U.S. Patent 4,176,385 (27 Nov. 1979).

24. J. Salge, U. Braunsberger, and U. Schwarz, "Circuit Breaker for Ohmic Heating Systems," *Proceedings of the Sixth Symposium on Engineering Problems of Fusion Research,* San Diego, CA, IEEE Publ. 75CH1097-5-NPS, p. 643 (1975).

25. Explosive generators are described in several papers in P. J. Turchi, ed., *Megagauss Physics and Technology,* Plenum, New York, (1980). Use of such generators together with the fast opening switches is described, for example, by T. J. Tucker, D. L. Hanson, and E. C. Cnair, "Experimental Investigation of an Explosive-Driven Pulse Power System," *Di-*

gest of Technical Papers of the Fourth IEEE Pulsed Power Conference, IEEE Cat. No. 83CH1908-3, p. 794 (1983).

26. Design and characteristics of the submicrosecond opening switch, originally patented by A. I. Pavlovskii (Ref. 23a) has been modified by others. For example, see J. H. Goforth and R. S. Caird, "Experimental Investigation of Explosively-Driven Plasma Compression Opening Switches," *Digest of Technical Papers of the Fourth IEEE Pulsed Power Conference,* IEEE Cat. No. 83CH1908-3, p. 786 (1983).

27. V. A. Glukhikh, O. A. Gusev, A. I. Kostenko, B. A. Larionov, N. A. Monoszon, M. A. Stolov, and G. V. Trokhachev, "Pulsed Sources of Energy Based on Inductive Storage," D. V. Efremov Research Institute, Report B-0299, Leningrad 1976.

28. Voltage hold-off exceeding 100 kV was reported by R. D. Ford and I. M. Vitkovitsky, "Explosively Actuated 100 kA Opening Switch for High Voltage Applications," NRL Memorandum Report 3561, Naval Research Laboratory, Washington, DC (1977). Much higher voltage (exceeding 600 kV) was obtained later and reported by I. M. Vitkovitsky, D. Conte, R. D. Ford, and W. H. Lupton, "Current Interruption in Inductive Systems with Inertial Current Source," NRL Memorandum Report 4168 (1980), as well as under the same title in *Energy Storage, Compression and Switching,* Vol. 2, V. Nardi, H. Salin, and W. H. Bostick, eds. Plenum, New York, pp. 953–971, (1983).

29. R. D. Ford and I. M. Vitkovitsky, "High Recovery Voltage Switch for Interruption of Large Currents," *Rev. Sci. Inst.* **53**, 1098–1100 (1982).

30. P. S. Levi, J. D. Watson, M. L. Hines, and R. E. Reinovsky, "Staged Explosively-Driven Opening Switch Development for Explosive Flux Compression Generators," *Digest of Technical Papers of the Fourth IEEE Pulsed Power Conference,* Albuquerque, NM, IEEE Cat. No. 83CH1908-3, pp. 782–789 (1983).

31. D. Conte, R. D. Ford, W. H. Lupton, and I. M. Vitkovitsky, "TRIDENT—A Megavolt Pulse Generator Using Inductive Energy Storage," *Digest of Technical Papers of the Second IEEE Pulsed Power Conference,* Lubbock, TX, IEEE Cat. No. 79CH1505-7, pp. 276–283 (1979).

32. A. E. Voytenko, V. I. Zherebenko, I. D. Zacharenko, V. P. Isakov, and V. A. Faleyev, *Fizika gorenia i vzryva* (Phys. of Combustion and Explosions) **11**, 145 (1974) (in Russian).

33. E. A. Azizov, N. A. Akhmerov, and V. A. Yagnov, "Influence of Dielectric Medium on the Characteristics of a High Speed Explosive Circuit Breaker," *Sov. Tech. Phys. Lett.* **2**, 121, (1976).

34. C. H. Johansson, *Detonics of High Explosives,* Academic Press, New York (1970).

35. R. D. Ford, D. J. Jenkins, and P. J. Turchi, *Pulsed High Pressure Gas Generator for the LINUS-O System,* Naval Research Laboratory Memorandum Report No. 3537 (1977).

36. E. A. Azizov, N. A. Akhmerov and V. A. Godonyuk, "Investigation of Effectiveness of Gas Flow Action on Arc Discharge Initiated in Contact Breaking," *Appl. Phys. Lett.* **45**, 232 (1984).

37. R. B. McCann, H. H. Woodson, and T. Mukutmoni, "Inductive Energy Storage Using High Voltage Vacuum Circuit Breaker," *Energy Storage Compression and Switching,* pp. 491–496, Plenum, New York (1976). Also in *Proceedings of First IEEE International Pulsed Power Conference,* Lubbock, TX, IEEE Catalog No. 76H1147-8 REG-5, (1976).

38. C. A. Bleys, D. Lebely, C. Rioux, and E. Rioux-Damidau, "Fast Circuit Breakers for 200 kA Currents," *Proceedings of International Conference on Energy Storage, Compression and Switching,* Torino, Italy, pp. 497–500 (1974).

39. V. L. Korolkov, M. A. Melnikov, and A. P. Tsyplenko, "Dielectric Breakdown of Detonation Products," *Sov. Phys. Tech. Phys.* **19**, 1569–1570 (1975).

40. Recent reporting of the work on opening switches in conjunction with explosively-driven current generators can be found in J. H. Goforth and R. S. Caird, Ref. 26 above; and in T. J. Tucker, D. L. Hanson, and E. C. Cnare, Ref. 25 above.

8
Fuses

INTRODUCTION

Metal fuses, sometimes referred to as "exploding wires" or "exploding foils," are used mainly for current interruption, i.e., as opening switches for circuit protection or in the generation of inductive voltage pulses. They are relatively simple devices, with a capability to perform demanding functions in applications of inductive storage technology. Fuses operate as a consequence of overheating by the current which they carry, resulting in their vaporization that causes their resistance to increase drastically and, effectively, cut off the current. They are employed in small bench-top generators[1, 2] including generators with very fast (100 ns) output voltage rise time[1] and in large, multi-megajoule systems[3, 4, 5] where they provide opening times as short as 1 μs. Very high voltages (6 MV) have been generated across an open circuit with a relatively small (134 kJ) Marx generator initially charged to about 1 MV. This device[2] has been used to produce 3.2 MV, 45 kA electron beams of short duration. As claimed by the authors of Ref. 2, this device is extremely inexpensive in comparison with high voltage Marx capacitor banks. Indeed, frequently fuse opening switches are employed to lower costs of laboratory pulse power devices.

To extend the usefulness of fuses (and to eliminate the need for Marx generators altogether) staging of fuses has been employed to provide power amplification up to the sub-terawatt level[6, 7]. Testing of power multiplication techniques using fuses in a very large capacitive energy storage systems (9.5MJ), with the short circuit current of almost 90MA,[4] recently has led to a significant improvement of the current interruption techniques discussed in this chapter.

In addition to the application of fuses in pulsed inductive energy storage systems, many studies of the exploding conductors have been motivated by other technology needs. For example, they are used to commutate the current away from mechanically operated circuit breakers (such as those discussed in Chapter 7) to provide an interval of time necessary for the breaker to recover[8]. They also are used to isolate sections of capacitor banks to protect them against destructive malfunctions, such as high current faults, that divert higher current to such sections[9]. Fig. 8-1 is a photograph of the 3 MJ capacitor bank in the process of assembly, showing each 22 kJ capacitor container (a steel can) with a cylindrical protective fuse package, available commercially[9].

The properties of fuses have been studied over a broad range of parameters and in a variety of circuit configurations. While these properties can be divided into thermodynamic, electrical, and mechanical categories, often the *interrelation* among these properties dominates the design of fuse switches. Thermodynamic properties include vaporization time (i.e., the time it takes the current to ohmically heat the fuse to vaporization and the

Fig. 8-1. This 3.5MJ capacitor bank, undergoing assembly at the Naval Research Laboratory, uses protective fuses mounted in cylindrical housings on each capacitor. Courtesy of J. Golden of Naval Research Laboratory.

consequent increase in resistivity) and the energy required to produce significant resistivity change. Electrical properties include resistivity as a function of time and current density, in particular the maximum achievable resistivity, the rate of change of resistivity, and the electric field that can be generated by the fuse and breakdown conditions, i.e., conditions leading to a drastic drop in resistivity either through development of a breakdown arc in the vapor or an overheating of the vapor that leads to ionization. Associated with breakdown is the so-called switch recovery rate (i.e., the increase in the amplitude of the breakdown field that can be sustained across the arc as a function of time between current interruption by the fuse and the time that the field can be reapplied again without significant current conduction through the switch), analogous to that discussed in conjunction with circuit breaker performance in Chapter 7. Recovery of other opening switches is also discussed elsewhere in this book, for example, in Chapter 9.

All of the electrical properties depend to some degree on the medium surrounding the fuse. Reference 10 describes drastic changes in the electrical properties of fuses as these are vaporized in water, air, or a vacuum. Finally, it is often necessary to consider the mechanical forces invoked by the rapid vaporization of the fuse or by the magnetic fields generated by the fuse current. Because the deposition of energy into the fuse is very rapid and energy density may be typically in 10^{10} J/kg, fuses switching larger circuits behave in a manner similar to activated explosives. Larger (multi-megajoule) energy storage systems switched by the fuse could deposit some few MJ of electrical energy into the fuse. Since 1 kg of chemical explosive releases roughly 4 MJ of energy, it can be seen how destructive the switching using fuses could be. Switching of large current could also lead to magnetic field forces acting on the fuse as it conducts. Such forces are sufficient to tear up the foil fuses or slam wire fuses against container walls, introducing unpredictable behavior, sometimes with disastrous results.

Exploding conductor fuses, used as opening switches, also have properties of importance in system applications. First, in contrast to closing switches, parallel fuse currents self-adjust so that modular approaches provide useful flexibility for the system design. For example, in systems with stored energy of hundreds of kilojoules and more, the paralleling of fuses provides a means for distributing the explosive force in a controlled manner, so that the pulser structural design can become simpler. Paralleling of fuses also helps to accelerate the rate of the resistance increase[11] to provide a faster opening time in comparison to that with fewer parallel wires. Fuses can also be operated with parallel resistors to help in sharing the energy that must be absorbed from the circuit (in cases where loads with inductive components are used). Lastly, the fuse has inductance associated with the finite length of the fuse, which is determined by the voltage that can be generated across a unit length of the fuse, as well as due to packaging constraints. This inductance can assume a dominant role in applications requiring generation of very high power ($>10^{11}$ W) pulses.

THE FUSE AS A CIRCUIT SWITCHING ELEMENT

Unlike at lower powers, where fuses are used for protection or in assisting the circuit breakers to function properly, in high power applications fuses are used as one of the means of interrupting the current in the inductor, thus leading to the formation of high inductive voltages and the subsequent transfer of current, characterized by a fast risetime, into a load. This can be viewed effectively as power multiplication relative to that which could be delivered directly from the current source to the load.

Fig. 8-2 illustrates switching of current from the inductor, L_0, storing the energy in a magnetic field generated by current i_0, to a load element consisting of a resistive and inductive component in the example shown. The current source, not further specified, can be, for instance, coupled inductively to L_0 or it can be provided by the insertion of a charged capacitor into the first loop seen in the circuit of Fig. 8-2. This circuit is simplified. It illustrates only the main features of the fuse switching and does not show any inductance associated with the fuse represented by the resistor R, nor any stray capacitive coupling, which may be important for *very high* power generation, where fast switching is required. Similar circuits, of course, are employed with other types of opening switches, such as those described in Chapter 7 and 9.

To gain an appreciation of the role of the fuse in inductive circuits, consider the simple case of Fig. 8-2, where a capacitor, C, serves as a current

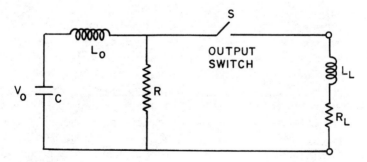

Fig. 8-2. Inductive circuit using a capacitor C as a current source to energize the inductor L_0. The resistor R represents time varying resistance having initial value R_i and final value R_F.

source to charge the inductor, L_0. Following J. Benford et al[12], if it is assumed that the initial fuse resistance is $R_i = 0$, the ratio of energy transferred from the capacitor to inductor has a value f (for the capacitor voltage, V_i):

$$f = (\tfrac{1}{2} L_0 \, i_0^2)/(\tfrac{1}{2} CV_i^2) \qquad (8\text{-}1)$$

so that

$$i_0 = \sqrt{fC/L_0} \cdot V_i \qquad (8\text{-}2)$$

If at some time during the energy transfer to the inductor, the fuse resistance, R_i, increases instantaneously to a value R_F, then the output voltage across the R_F becomes

$$V = R_F i_0 \qquad (8\text{-}3)$$

before the output switch, S, is closed. The ratio of the inductively generated voltage, V, to the capacitor charging voltage, V_i, i.e., the voltage multiplication factor is:

$$V/V_i = R_F \sqrt{fC/L_0} \qquad (8\text{-}4)$$

For $R_i = 0$, energy transfer time between C and L_0, τ, for the circuit in Fig. 8-2 is

$$\tau = (\pi/2) \sqrt{L_0 C} \qquad (8\text{-}5)$$

If at peak voltage appearing across R, generated by the change in R from R_i to R_F, the output switch is closed, the discharge time of the inductor L_0 (assuming $L_0 >> L_L$), defined as an e-folding time, Δt, is (assuming the value of the load resistance shown in Fig. 8-2 is $R_L = 0$):

$$\Delta t = L_0/R_F \qquad (8\text{-}6)$$

Thus the ratio of inductor charging time to the output pulse (in the circumstances when the time for the fuse resistance to change can be neglected in relation to the pulse duration determined by the circuit parameters), i.e., the pulse compression ratio, is:

$$\tau/\Delta t = (\pi/2) \, V/V_i \qquad (8\text{-}7)$$

In the above example, the fuse resistance behaves in a simple manner. In reality, there are many parameters which characterize fuse performance. Two of these, the maximum voltage (proportional to fuse length) which the fuse can maintain and its rate of change of resistivity, lead to energy losses in the fuse and determine its ability to transfer the current of the inductor to the load. Similar issues are addressed rigorously in the next chapter in conjunction with opening switches based on plasma conductors. The crucial difference between fuses and some plasma devices (e.g., those which use electron beams to control switch conductivity as discussed in Chapter 9) is that, while the plasma devices depend on control from external sources of energy, fuses operate (change their resistance) as a result of absorbing the energy from the circuit which is being switched. In the case of the circuit shown in Fig. 8-2, some of the energy stored in the inductor, L_0, must be reserved for fuse vaporization with the remainder being available for the load.

The energy transfer to the load, W_L, has been derived in Chapter 2 to illustrate the general property of inductive storage circuits, where opening switches always absorb some energy as long as they serve to transfer current into a circuit branch with an inductive component. In the case of the circuit in Fig. 8-2, that particular inductance is L_L.

The transfer time, t_T, for arbitrary values of the load resistance, R_L, and inductance, L_L, is derived following Ref. 13. After the series switch, S, in Fig. 8-2, closes, so that the voltage is applied across the load, the current through the load is determined by the equation

$$\frac{di_L}{dt} + \frac{i_L R_L}{L_L} = V(t)/L_L \tag{8-8}$$

Commutation of the current from the switch to the load is considered completed when the current in the storage inductor, L_0, flows in the load, i.e., the commutation time corresponds to the condition $i_L(t_T) = i_0$ (where i_0 is the current in the storage inductor) and $i_1 = 0$. The simplest case to analyze, and the one that most clearly illustrates what is significant in *minimizing* the transfer time, is that in which the fuse has developed maximum voltage, V_{max}, before the switch, S, is closed, so that $V(t) = V_{max}$ in Eq. 8-8. The solution to Eq. 8-8 becomes

$$i_L(t) = \frac{V_{max}}{R_L}[1 - e^{-at}] \tag{8-9}$$

where $a = R_L/L_L$. This solution shows that $i_L(t) \leq V_{max}/R_L$ so that full commutation of the current can be achieved only if

$$R_L < V_{max}/i_L \tag{8-10}$$

This condition must be satisfied in all cases, since V_{max} is the greatest possible voltage available to drive the load current, which can never exceed V_{max}/R_L.

Commutation will be achieved at some time if the above condition on load resistance is satisfied. The current transfer, $i_0 = i_L$, occurs at a time given by

$$t_T = \frac{1}{a} \ell n \left(\frac{V_{max}}{V_{max} - R_L i_0} \right). \tag{8-11}$$

In the limit, as $R_L \to 0$, the build-up of the load current is limited solely by the inductance of the load:

$$\frac{di_L}{dt} = \frac{V_{max}}{L_L} \tag{8-12}$$

so that

$$i_L(t) = \frac{V_{max}}{L_L} t \tag{8-13}$$

and the commutation time is

$$t_T = L_L i_0 / V_{max} \tag{8-14}$$

This is the shortest possible time for current transfer for a given load inductance. If both the resistance and the inductance of the load are finite, then t_T increases as R_L increases, as shown in Fig. 8-3. This figure is a plot of t_T given by Eq. (8-11) in units of $L_L i_0 / V_{max}$ as a function of load resistance expressed by the normalized variable $R_L i_0 / V_{max}$. The graph shows that for resistance values just slightly less than the critical value required for commutation, V_{max}/i_0, there is commutation, but it takes a long time. As resistance is reduced, the commutation time decreases rapidly toward commutation time for $R_L = 0$. When the resistance is half of the critical value, the commutation time is only 37% greater than the value for the $R_L = 0$ case.

The inductor current after commutation to the load will decay independently of the fuse if the dissipative element $R_L \ll R_F$. The decay time characterizes the time for the energy stored in the inductor, L_0, to dissipate in the load, R_L. This decay time can be derived from the voltage equation

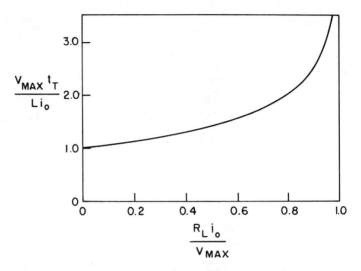

Fig. 8-3. Effect of load resistance in the circuit shown in Fig. 8-2 on the rate of current transfer from the exploding switch to the foil fuse. Reproduced from Ref. 13.

for the circuit in Fig. 8-2 after switch S has closed. Assuming the fuse resistance, R_F, at that time such that $R_F >> R_L$:

$$L_0 \frac{di_L}{dt} + L_L \frac{di_L}{dt} + i_L R_L = 0 \qquad (8\text{-}15)$$

The solution for the current in the circuit is

$$i_L(t) = i_L(0)\, e^{-\alpha t} \qquad (8\text{-}16)$$

where $\alpha = R_L(L_0 + L_L)$ is the characteristic decay time of the current.

The final value of the resistance achieved during the vaporization of the fuse plays a more crucial role when such fuses are employed in high power circuits where an inductor is used for charging a capacitor. Such a capacitor may be constructed as a pulse-forming line for appropriately shaping the output pulse[11, 14]. Inductive switching for charging of capacitors or pulse lines is shown in Fig. 8-4. Similarly to the case considered in Fig. 8-2, the current in the inductor, i_0, flows initially in the fuse. Vaporization of the fuse interrupts the current, generating a voltage pulse across the capacitor, C, provided the switch, S, is closed sometime during vaporization. If the fuse resistance, R_F, increases to infinite values, the energy remaining in the inductor will slosh back and forth indefinitely between the inductor and

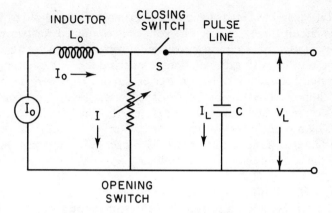

Fig. 8-4. Inductive storage for pulse line charging. The pulse line is represented as a capacitor C.

the capacitor. However, in reality, the finite value of the resistance, R_F, will absorb the energy in characteristic time $R_F C$.

To help in the selection of the fuse parameters necessary to charge a capacitor using inductively stored energy, as in Fig. 8-4, Fig. 8-5 should be used. Normalized voltage, $(V/I_0) \sqrt{C/L_0}$, is computed for aluminun fuses exploded in water using the experimental results summarized in Ref. 19. It is plotted in Fig. 8-5 (solid lines) as a function of the characteristic switch

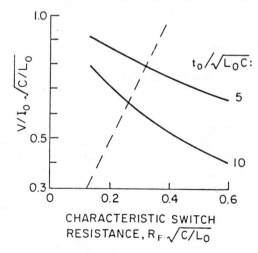

Fig. 8-5. Charging voltage on the pulse line as function of normalized switch resistance, R_F.

resistance, R_F, as measured at vaporization time, t_0 (normalized by the characteristic circuit time, $\sqrt{L_0C}$). The voltage is computed for two values of t_0, representative of a typical circuit employing an inductor to charge a capacitor, so that with extrapolations a switch designer can estimate what the charging voltage might be for a given circuit. However, the capacitor voltage cannot exceed the product of the electric field, E (at which value restrike occurs in the fuse), and the fuse length. The limit is represented in Fig. 8–5 by a dashed line, $E\sqrt{t_0} = 2600$ V/m $-$ s$^{1/2}$, conservatively taken in accordance with the results of Ref. 17. For a short vaporization time, $t_0 \simeq 4$ μs, the constant can be at least 4000 V/m $-$ s$^{1/2}$ according to the data in Ref. 19.

Fig. 8–5, in addition to estimates of voltage that can be developed across a capacitor (pulse line), also provides an estimate of efficiency of energy transfer from the inductor to the capacitor, with an account of the energy dissipated in the fuse. The efficiency is $\eta = CV^2/L_0I_0^2$, obtained by squaring the normalized voltage plotted in Fig. 8–5.

DOMINANT SWITCHING MECHANISMS

The rapid increase in the resistance of the fuse has been exploited to interrupt current. It results from Joule heating of the initially conducting metal carrying the current. If the heating is sufficient to obtain high temperature, a large ratio of R_{max}/R_0 can be achieved. Typically this ratio of the maximum resistance of the fuse, R_{max}, and the initial resistance at room temperature, R_0, is of the order of 100. For most cases $R_{max} = R_F$ and $R_0 = R_i$ defined earlier. To maximize this ratio, the heating rate must be sufficient to melt the fuse and vaporize it. An upper limit on the temperature is the incipient ionization, which leads to avalanching of free electrons in the vapor as they are driven by the inductively or resistively generated potential across the fuse. This rather delicate operation of a fuse which must be heated to vaporization, yet kept below the temperature that leads to ionization, is the main point, for example, of studies such as that of Ref. 15 and 16.

To obtain the optimum operation of the fuse, a substantial amount of empirical data has been developed and used in conjunction with analytical expressions to derive scaling laws that can predict the fuse behavior for given circuit conditions. For a uniform current distribution in the fuse in a solid or liquid state (that is, when the skin effect can be neglected) when thermal loss to the electrodes and to the surrounding medium can be neglected, and dynamic instabilities are avoided, the heating rate of the fuse is governed by Joule heating and has a simple expression:

$$m\frac{dW}{dt} = R_F I^2 \qquad (8\text{–}17)$$

Here, W is the internal energy per unit mass and the fuse resistance, R_F, can be expressed in terms of resistivity ρ:

$$R_F = \rho(h/s) \qquad (8-18)$$

where s is the fuse cross-sectional area, h is its height, and m its mass and C_m is the specific heat of the fuse. Such heating leads to the rise in temperature, T, in a fuse of mass m according to the relationship

$$C_m \frac{dT}{dt} = \frac{dW}{dt} \qquad (8-19)$$

The rate of increase of temperature or internal energy can be determined numerically from given $I(t)$ and from measuring $\rho(t)$, assuming that s does not change during the process.

In practical applications of these relations, the current needed to vaporize the fuse can be reduced to analytic form. For example, Maisonnier et al.[15] consider a capacitor (with capacity C_0) charging an inductor, L_0, (with a sinusoidal waveform) and calculate the condition for the vaporization (explosion) of the fuse at maximum current, I_0, at a time $(\pi/2 \sqrt{L_0C_0})$. They derive the condition for the explosion to be given in terms of the indictive energy, $W_0 = \frac{1}{2}L_0I_0^2$ and capacitor charging voltage V_0:

$$W_0^{3/2}/(V_0 L_0^{1/2} s^2) = k_1 a \qquad (8-20)$$

The constant, k_1, with a value ranging from 1 to 3, is needed to account for a variety of factors that are not included in the simplified analysis. The constant a is defined as

$$a = (\sqrt{2}/\pi) \, \gamma \int_{\text{init. temp.}}^{\text{vapor. temp.}} \rho^{-1} \, dW \qquad (8-21)$$

This constant thus contains the remaining physical parameters, with γ designating the mass density of the fuse.

Resistivity

To help in estimating the behavior of a fuse in a given circuit, ρ has been measured for a variety of fuse materials and for various media surrounding the fuse, usually as a function of the internal energy. The resistivity of copper before melting behaves linearly with the temperature, T:

$$\rho = 1.58 \times 10^{-8} (1 + 5.15 \times 10^{-3} T) \text{ } \Omega\text{-cm} \qquad (8\text{-}22)$$

where T is in °K. The resistivities at room temperature (20°C) and the melting point temperature (1083°C) are $1.74 \times 10^{-8} \Omega$ − cm and 1.04×10^{-7} Ω − cm. The energy density corresponding to the melting point is 4.17×10^{9} J/m³. These parameters suggest that, in most calculations, the effect of the fuse on the circuit can be neglected. A set of measurements is also given in Ref. 15 with temperature indicated along the ρ^{-1} vs. W curve for aluminum. W. H. Lupton[16] has pointed out that the intrinsic form of Eq. 8-17,

$$j^2\rho = \frac{dw}{dt} \qquad (8\text{-}23)$$

where j is the current density and w is the energy *per unit volume,* implies

$$\int j^2 \, dt = \int \frac{dw}{\rho} \qquad (8\text{-}24)$$

The left-hand side of this equation is the so-called "action" integral[15]. The right-hand side is a function of the material properties only and therefore for $\rho = \rho(w)$,

$$\int_0^\infty j^2 \, dt = \frac{\pi}{\sqrt{2}} a \qquad (8\text{-}25)$$

The values of the constant a are given in Ref. 15. The values, in mks units, are 3.9×10^{16} for Ag, Au, and Al, respectively. (The authors of Ref. 15 chose to use energy per unit mass, rather then energy per unit volume used in this discussion.) It should thus be possible to check experimentally whether for constant current density, j, the scaling for the time to vaporization obtained from Eq. 8-25,

$$j = 1.49 \sqrt{\frac{a}{t}} \qquad (8\text{-}26)$$

is valid.

Braunsberger, Salge, and Schwarz[17] used copper wires to demonstrate, as shown in Fig. 8-6 (reproduced from their work), that

$$j = j_0 (10^6 t)^{-0.54} \qquad (8\text{-}27)$$

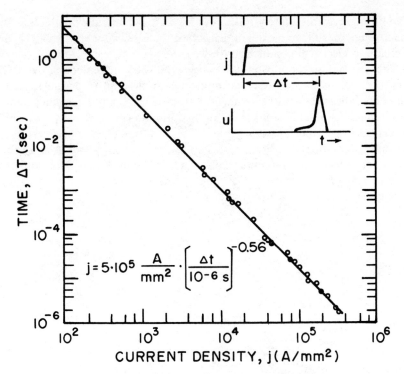

Fig. 8–6. Time-to-explosion of copper fuses versus current density in the fuse. The current waveform in this experiment is shown in the insert. Reproduced from Ref. 17.

where $j_0 = 5 \times 10^7$ A/cm^2 and t is in seconds. The time from current start to the explosion given by this scaling is independent of length and cross-sectional area of the fuse, provided the current is constant. This experimentally measured power law (Eq. 8–27) over many orders of magnitude is very close to that given on the basis of simple considerations used in deriving the relationship (8–26).

The fact that the time-to-explosion of a fuse is proportional to current density is the basis presented by Maissonier et al.[15] for determining the optimum fuse cross-sectional area. It is presumed that in designing an opening switch it is desired to adjust this time to result in the maximum magnetic energy stored in the inductor. In Ref. 15, the source of current was a capacitor, so that the time to peak current was determined by the capacity of the primary storage and the inductance of the magnetic storage; the criteria for selecting the fuse cross-section was obtained by equating the current risetime to the time-to-explosion of the fuse. With the cross-section thus determined, the fuse length is calculated by considering energy dissipation.

Too little energy dissipated in the fuse leaves it as a metalic conductor. Too much energy results in partial ionization of the fuse vapor with an attendant increase in electrical conductivity[15, 16].

The above design philosophy has been followed and extended by W. H. Lupton[18] for use with numerical models in an attempt to design fuses for more complicated circuits in which the fuse energy is not determined simply by circuit constants. He makes use of an optimum energy density, w_0, which is determined experimentally. Fig. 8-7 is a model of energy density-dependent conductivity. Such models must be used with care because their exact form depends on many factors, such as the medium surrounding the fuse or rate of energy deposition into the fuse. For example, the effect of energy deposition is discussed later in conjunction with Fig. 8-8.

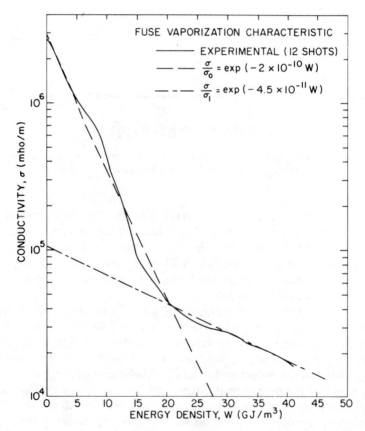

Fig. 8-7. Al fuse conductivity (normalized to room temperature conductivity) shown as function of the energy density deposited into the fuse.

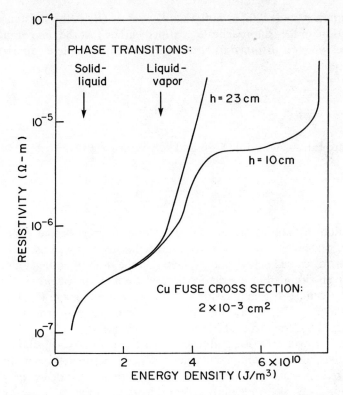

Fig. 8-8. Resistivity dependence of copper fuse on the rate of energy deposition into the fuse.

An optimum fuse length will result in an energy density w_0, just sufficient to vaporize the fuse. In the application addressed by Maissonier et al.[15], the fuse switches the current from one inductor to another, and the energy to be dissipated is precisely determined by circuit inductances and current. Lupton[18] calculates the fuse length by also equating this dissipated energy to the product of fuse volume, hs, and the optimum energy density:

$$w_o hs = \int_0^\infty IV dt \qquad (8\text{-}28)$$

This expression for purposes of dimensional analysis can be simplified by assuming a constant current source and some average voltage across the fuse, V_{av}, during the fuse conduction period. Then

$$hs = (IV_{av} \, t/w_0) \qquad (8\text{-}29)$$

To eliminate the length, s, note that $V = R_V I$. (R_V is the resistance of vaporization and R_V approaches R_F defined earlier.) At the time of maximum voltage, which is *assumed* to occur at the time of complete vaporization of the fuse,

$$R_F = \frac{h}{s} \rho_0(w) = V_{max}/I \qquad (8\text{-}30)$$

Dividing Eq. 8-29 by Eq. 8-30 and solving for t gives

$$t = \frac{s^2}{I^2} \left(\frac{w_0}{\rho_0} \right) \left(\frac{V_{max}}{V_{av}} \right). \qquad (8\text{-}31)$$

The values w_0 and ρ_0 are measured material properties that depend, to a degree, on the environment surrounding the fuse, as well as on other parameters discussed below, and explicitly on the ratio of voltages determined by a given circuit. For a specific case, where a capacitor provides the current, V_{av} can be written in terms of initial capacitor voltage V_c : $V_{av} = \alpha V_c$. The ratio V_{max}/V_{av} has a value, typically, of 4 to 6 as seen from the example considered by Benford et al.[12], and α is in a range of 1 to 2.

Considering a specific example of Al fuses, the ratio of w_0 and ρ_0, can be evaluated from the conductivity dependence measured for aluminum foil vaporized in water[5, 19]. Figure 8-7 shows that dependence graphically. Equation 8-32, below, is derived from the data in Fig. 8-7, to provide the resistivity in analytic form (representing the dashed lines). (In Fig. 8-7, the conductivity plot starts at the onset of vaporization, so that the absolute energy density scale is offset by a small amount required to bring the fuse to vaporization). The analytic form of the resistivity is in Ω-m:

$$\begin{aligned} \rho &= 8.6 \times 10^{-8} \exp(-2 \times 10^{-10} \, w) \text{ for } w < 2.8 \times 10^{10} \text{ J/m}^3 \\ &= 6.6 \times 10^{-6} \exp(-4.5 \times 10^{-11} \, w) \text{ for } w > 2.8 \times 10^{10} \text{ J/m}^3 \end{aligned} \qquad (8\text{-}32)$$

Taking $w_0 = 3.3 \times 10^{10}$ J/m^3 for Al and the value of ρ at this energy density of 3×10^{-5} Ω-m, and the $V_{max}/V_{av} = 6$, Eq. 8-31 becomes: $t_{Al} = 7 \times 10^{15}$ s^2/I^2. Note that the value of t for Al is within a factor of 3, in agreement with the scaling in Eq. 8-26. The numerical coefficient evaluated in this manner is quite approximate, analogous to the range of values assigned to k_1, mentioned in conjunction with Eq. 8-20. Such an evaluation serves only as a guide in determining the cross-sectional area, s, needed for a given current in the fuse.

A study of the resistivity of the fuse vapor column in air is given by L. Vermij[20]. It includes a calculation of the resistivity from measured experimental parameters as a function of the (silver) vapor temperature in a regime where thermal ionization takes place. In the temperature regime for 3,000°K to 16,000°K, resistivity decreases by five orders of magnitude.

The presently fragmentary knowledge of the physical phenomena describing the resistivity in various fuse configurations makes it difficult to extrapolate the performance of a fuse to regimes where no performance data are available. In addition, better measurements as well as a more complete understanding of the vaporization phenomena will likely lead to the development of fuses with much better electrical performance in their role as opening switches.

An example of the difficulties in extrapolation is the dependence of resistivity on the energy density in the fuse, as already suggested above. This dependence is not a unique function for any given metal. One reason for this is the tacit assumption that the cross-sectional area of the fuse does not change (i.e., it is assumed that $ds/dt = 0$) with heating, resulting in definition for the resistivity used above that incorporates the ratio of the initial fuse length, h, and cross-sectional area, s. (J. D. Logan et al.[21] discuss, with referencing to other work, the difficulty of treating the conductivity of fuses that change their volume during heating.) It can be expected, however, that at least during vaporization, the rate of input energy may determine the expansion of the fuse and therefore its resistance, as well as its resistivity as defined above. This was initially noticed when fuse switches were used to transfer the energy stored in an inductor that was energized by a long current pulse from a homopolar generator[22]. Since the various values of shunts were used in parallel with the fuse (to reduce the explosive energy of the fuse without reducing the output voltage), the function $\rho(w)$ assumed different values.

For a better understanding of the behavior of resistivity as the rate of energy dissipated in the fuse varies, a series of experiments has been performed by W. H. Lupton and R. D. Ford[23]. Figure 8–8 shows the variation of resistance with the deposited energy density for two different lengths of the fuse. When the current density and waveform are kept the same in both cases, the rate of energy deposition differs by a factor of 2.3 between the two fuses. The longer fuse, with a lower energy deposition rate, increases more rapidly in resistance. The difference in $\rho(w)$ does not appear until sufficient energy to initiate vaporization has been delivered to the fuse.

Logan et al.[21] have gone beyond simple scaling relations to develop a method of computing the transient current distribution and ohmic heating of two-dimensional electrically exploded conductors before vaporization. Because they consider confined fuses, their treatment of the vaporization phase differs from earlier work, which postulates that the resistance spike during the explosion is due to blow-off of the vaporizing material from the wire (causing a corresponding reduction in the conductor cross-section). With their analysis limited to short heating times ($\lesssim$ 1 μs), their model approximates a two-dimensional case quite well. Their assumptions appear to make the computation method applicable to fuse explosions confined by fuse packaging such as that discussed below. One interesting result of the analysis in Ref. 21 is that observation of the temperature distribution in foil fuses shortly before an explosion makes it possible to predict the degree of simultaneity of the heating of all sections of the foil. This suggests that such calculations may guide in developing fuse geometries that maximize the rate of resistance increase and result in higher final resistance, as well as lead to higher hold-off fields.

Geometric Effects

In addition to cross-sectional area, s, and length, h, there is a host of other effects of fuse geometry that should be considered in designing such switches. The significance of a particular aspect of fuse geometry on performance is, of course, determined by the particular application. Inattention to these factors may lead to a severe degradation of performance.

The first of the geometric factors to consider is fuse diameter or thickness in relation to electromagnetic field penetration (diffusion) length. If for a given current risetime in a fuse, the field does not penetrate fully into the fuse, heating will not be uniform so that some layers may become nonconducting while others remain conducting. At the least, this would lead to prolongation of the switch opening time. One of the first groups to obtain very fast opening times for fuse switches, that of Yu. Kotov et al.[24], demonstrated the need to subdivide a single fuse into an array of parallel fuses to avoid the effects of skin thickness on opening time. For example, they show the opening time of a single wire fuse (with a diameter of 0.28 mm) is twice as long (80 ns) as that of an array of 12 wires (each 0.08 mm in diameter). Conte et al.[19] have extended these studies to higher voltages ($\leq$ 900 kV). The results are reproduced in Fig. 8-9, together with a circuit used in the experiment and a typical open circuit voltage waveform. The waveform shows that the inductive voltage rise, initiated by the foil fuse, reaches about 200 kV, at which time a closing switch (over-voltaged gap) closes to commutate the current from the inductor to the wire array fuse. The explosion of that fuse by interrupting a large current multiplies the

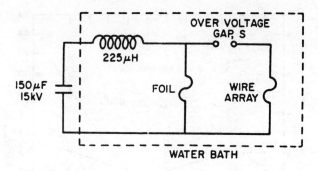

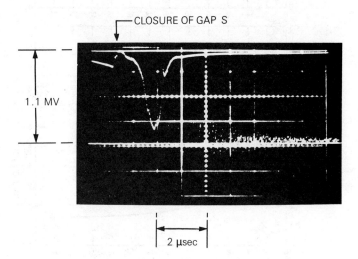

Fig. 8-9. Test circuit and typical voltage waveforms of staged exploding foil and wire fuses. Reproduced from Ref. 19. © 1977 IEEE. *Fig. continues on p. 188.*

inductively generated voltage to about 900 kV (i.e., a factor of 4 to 5 higher then the initial voltage).

While Fig. 8-9 shows the effect of increasing the number of wires in the fuse array, experiments demonstrating the effect of the diameter of a single wire fuse on the inductive voltage[23] generated by the interruption of current

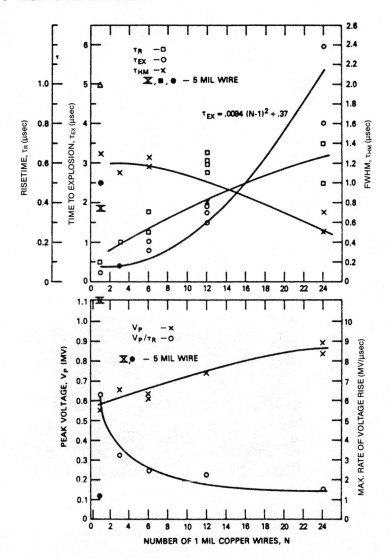

Fig. 8-9 (*cont.*) Risetime, pulse duration, time to explosion, and amplitude of voltage pulses generated by copper wire arrays in water (driven by an aluminum foil fuse) are shown. Reproduced from Ref. 19. © 1977 IEEE.

in the inductor is shown in Fig. 8-10 for time-to-vaporization of about 20 µs. For a change of cross-section area by a factor of 2, the ratio of voltages generated by the two different fuses increases mildly, from 1.36 for 10 cm fuses to 1.42 for 18 cm fuses. These experiments also demonstrate the dependence of the electric field generated by the fuse on the fuse length. At

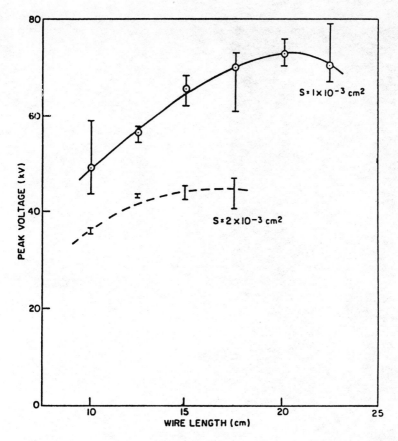

Fig. 8–10. Variation of the peak inductive voltage generated by a fuse as a function of its length.

shorter fuse lengths, the capacitor, serving as a source of current through the fuse, can deliver the current for a sufficiently long time to raise the energy density in the fuse to drive it to high resistance. In this regime, the inductive voltage generated across the high resistance of the fuse is proportional to the fuse length, i.e., the electric field is independent of the fuse length. As the length increases (beyond 16 cm for the parameters associated with Fig. 8–10), its mass increases to the point that the energy source cannot maintain the current through the fuse for a sufficient time. This results in a lower energy density in the fuse and consequently in lower final resistance, leading to inductive voltage decreasing at longer fuse lengths. This represents another limit, in addition to the restrike hold-off already mentioned, on the maximum voltage that can be generated inductively using fuses as opening switches.

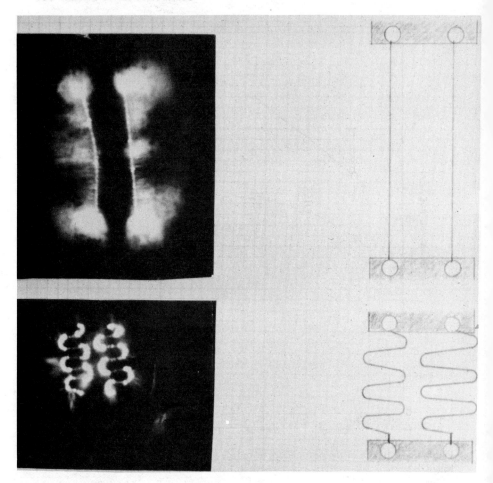

Fig. 8–11. Degree of uniformity of 9 cm long fuse expansion during switching as function of fuse configuration. Commercial copper wires, gauge No. 27, were used.

Bending of the fuse (as may be done to shorten the over-all length of a switch), and the resultant effects, is another factor that must be considered in evaluating a fuse as a switching element.[23] Increased brightness seen in time-integrated photographs of fuses vaporized in water, shown in Fig. 8–11, reflect the localized heating of each fuse at electrodes and at bends. These photographs are interpreted as follows: The distortion of metal wire by bending or clamping at electrodes leads to localized thinning of the wire. Parts of the fuse with smaller cross-sections vaporize earlier and continue to be heated, since the overall resistance of the fuse is not sufficient to

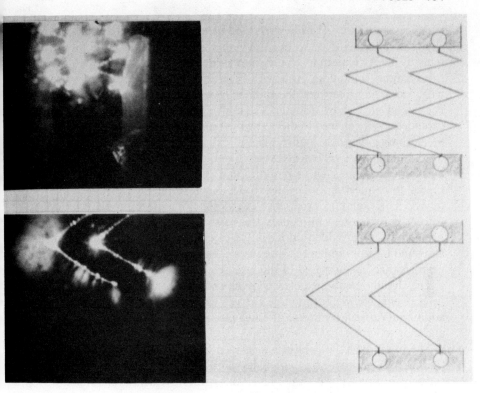

Fig. 8-11 (*cont.*)

interrupt the current from the capacitor employed as the current source. Continuing heating leads to ionization and reduces the resistance of the affected sections to insignificant values. The undistorted sections vaporize later, with the total fuse resistance and rate of rise of resistance reduced according to the relative lengths of cold and hot sections. The peak inductive voltage generated by the current interruption by each fuse configuration, shown in Fig. 8-11, differs by a factor of 2 between the best, straight-fuse case and the worst, multiple sharp-bend fuse configuration.

In Figure 8-11, straight sections of the fuse contain short bright and overheated regions. This can be attributed to non-uniformities in the diameter of the fuse made from commercially produced wires. It indicates that more uniform fuses would provide better performance.

In addition to wire fuses, metal foils also are available commercially in many thicknesses and widths (especially aluminum foils), providing a convenient material to be used for opening switches. Typically, foils used for switches range in thickness for 0.001 to 0.01 cm and are tens of centimeters wide. For example, Al foils were used as a first stage switch in the circuit shown in Fig. 8-9. Copper foils have been used in larger pulse power systems by R. A. Haarman et al.[25] in disposable packages, discussed later in this chapter. In analogy with wire fuses that exhibit more rapid switching with decreasing diameter, the thinner foils show a similar tendency. V. A. Burtsev et al.[26] show also that edge effects in high current foil Al fuses exploded in air can lead to dielectric breakdown of the gap and to formation of discharge channels before the foil expansion is completed. A point of diminishing returns is reached when handling of thin foils becomes difficult and imperfections such as pinholes and small tears at the edges, and difficulty in connecting at the electrodes begin to degrade the foils' performance. For example, experience shows that the cutting tools for making foil strips ultimately influences the formation of edge discharges. These problems have provided an incentive to consider the effects of the media surrounding foil (and wire) fuses as a means of improving fuses' electrical performance, as discussed in the next section.

The already-mentioned effects of geometry influence the performance of fuses in small as well as in large energy systems. In large systems, there are additional difficulties in use of fuses associated with strong hydrodynamic and magnetic pressure effects.

The hydrodynamic pressure generated by rapid vaporization of each of the wires in parallel arrays of fuses can lead to mechanical disruption of the neighboring wires or foils. The rarefaction wave behind the shock wave, after interception of the neighboring wires, also can degrade the hold-off characteristics of the affected element of the fuse array and therefore lower the hold-off capability of the entire array. Such cylindrical rarefaction waves (produced by laser-guided discharges) have been studied quantitatively by Greig et al.[27], who show the rarefaction of atmospheric air by a factor of 20 at 40 μs from burst of the fuse for a deposition energy of 300 J/m. Relaxation to initial conditions is slow, requiring some 10 ms to approach the undisturbed conditions. This result coincides with the observation of Borisov et al.[28], that the exploded fuse channel recovery field, E_r, discussed later in the chapter, increases at a faster rate of recovery (10^{-4} to 10^{-2} s range) in the case of air quenching. Reference 27 also shows that the rarefied channel reaches assymptotically a maximum diameter of about 2 cm between 50 to 100 μs from an initial velocity of about 0.1 cm/μs. These parameters indicate that an array of wires in a fuse operated in air need not be separated by more than 1 cm if the fuse's electrical characteristics resulting from reduced pressure environment are to remain unchanged from those of a single wire fuse. Although similar effects have been noted in fuses operating in water, no qualitative measurements are available.

The effects of magnetic pressure generated by current flowing in the neighboring fuse array elements or in a current-return electrode can produce mechanical disruption of the fuse element even before the hydrodynamic pressure effects can arise, degrading their electrical performance. This effect becomes significant at current levels of the order of 100 kA and fuse current flow duration of tens of microseconds.

Environmental Effects

The behavior of fuses described in the preceding sections can be modified by replacing air or other gases surrounding a fuse (quench medium) by solid or liquid materials to prevent fast expansion of the fuse products. Keeping the gaseous products at high pressure helps to maintain high dielectric strength of the fuse channel. In contrast, the vapor products of fuses exploded in evacuated containers expand very rapidly and exhibit a much different electrical performance as a switch.

Considerable research has been performed to determine the conditions required to maximize the induced electric fields that can be supported across the fuse, as well as maximize the recovery fields. The motivation for this work stems from the fact that the fuse used as an opening switch in pulse power systems must hold-off voltages as near to those appearing across the load as possible for optimum circuit efficiency. The dielectric strength of the fuse switch during vaporization and after the recovery period, as well as fuse resistivity, are functions of time and can be controlled to a substantial degree by the medium surrounding the fuse switch.

The performance of fuses in water[5, 17], in sand-like materials[26, 28, 29] (and recently in chemically-active gas[23], SF_6), as well as in a vacuum[31] has been studied adequately to provide useful data for switch design. The references provided here are not meant to be exhaustive, but rather they are intended to provide the reader with leads to additional literature. Fuses operating in water, quartz powder, and air have been used as switches in a variety of pulse power systems; the more significant examples of such systems are described in the references cited here. Fuses operating in a vacuum, while promising to eliminate those problems which result from the explosive pressure impulses associated with the use of quench media, have not yet been evaluated as switches in any pulse power systems.

Burtsev et al.[26] are concerned with the changes in the resistance, R_F, of copper foil fuses (normalized to the initial room temperature resistance, R_i) as a function of the deposited energy per unit mass, W, for several quenching media. They choose to characterize the effects of media on fuse resistance by specifying resistance values just before vaporization and at the beginning of vaporization. The following relations satisfy the empirical data:

$$R_F/R_i = 1 + \gamma \cdot W \quad \text{before vaporization}$$
$$= 20 + \delta \cdot \Delta W \quad \text{at vaporization}$$

(8-33)

Here ΔW is the difference between the energy required to melt and to vaporize a fuse depending on the quenching media. To the first order, $W = \Delta W$. The values of γ and δ are tabulated according to Ref. 26, with quenching media shown at left:

	$\gamma(g/kJ)$	$\delta(g/kJ)$
Air	6.0	35
Polyethelene	5.3	27
Paraffine	5.0	20
Water	4.5	15
Quartz dust	4.5	15

Thus, for example[26], it is necessary to add 4.5 kJ/g to fuses operating in water or quartz sand, compared with 3.2 kJ/g for the air-quenched fuses, in order to change the resistance ratio, R_F/R_i, by about a factor of 20.

Equation 8-33 and the tabulated values supplement the expression for the continuous change of resistance (with energy input) given by Eq. 8-32 by providing an estimate of the effect of quenching media on foil fuse performance in frequently used circuits where the conduction time is in the microsecond regime.

V. E. Scherrer and this author[10] have been concerned with the effects of the environment on the ability of the fuse to hold-off high electric fields that could appear across it due to interruption of current in an inductor, resulting from the fuse's vaporization and its ability to hold-off electric fields which appear upon their interruption by subsequent fuses, as shown in the circuit in Fig. 8-12. The staging of parallel opening switches* employed in Ref. 10 is a practical method for generating high power pulses from inductive storage systems charged at much lower power. Such staging has been used in several inductive storage pulsed system development programs[7, 17, 32]. (Ref. 32 describes yet another approach for using the quench media to control fuse parameters: an active explosive material is used to compress rather than merely confine the vaporization products.) It is also a convenient technique for generating high voltage pulses with different delay times relative to first fuse vaporization, useful for studying the restrike properties of the fuse.

*The concept of opening switch staging is introduced in Chapter 2.

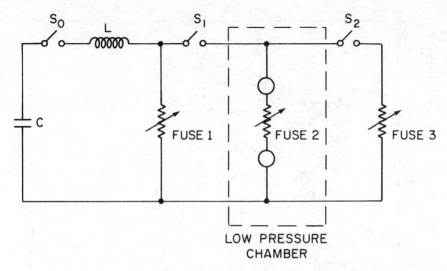

Fig. 8-12. Inductive circuit for determining restrike voltage of fuses in vacuum. Capacitor C (60 μF) charges up to 10 kV. Inductance L is approximately 3.5 μH. No lead and fuse inductances are shown. Reproduced from Ref. 10.

Fuses Operating in Air at Atmospheric and Low Pressures. Figure 8-13 shows the dependence of the restrike field, E_r, on t_e for a 0.12 mm diameter Cu fuse in a vacuum. For comparison, the same dependence of the restrike field of fuses exploded in air, as obtained by Braunsberger et al.[17] and in our study (at long and short charging times respectively) is shown. The values of E_r in a vacuum and in air (for $t_r \lesssim 100 \; \mu$s) were obtained using the circuit shown in a simplified form in Fig. 12. A low voltage capacitor was discharged through fuse 1, with time to peak current of 15 μs. By the selection of different fuse 1 diameters, t_e was varied (together with the current in the fuse at t_e) to provide the data in Fig. 8-13 for $t_e > 3 \; \mu$s. Data for $t_e < 3 \; \mu$s were obtained by using fuse 1 as a current-steepening switch. Commutation of the current to fuse 2 provided up to 10 kA, with a rise of 1.5 μs, into the low pressure chamber. The variation of the diameter of fuse 2 provided the data points for $0.1 < t_e < 3 \; \mu$s.

Although E_r of up to 10 kV/cm can be generated by fuses in a vacuum, this field does not exceed that obtainable with fuses when their expansion is partly confined by air at atmospheric pressure. Figure 8-14 shows how E_r of Cu fuses varies with the pressure of the surrounding air. The confining function of the surrounding medium is negligible for $P \lesssim 50$ torr. Furthermore, the restrike field is always less than the electrode gap breakdown field, E_b, in the pressure range of 10^{-2} to 50 torr; $E_r \simeq E_b$, with E_r rising

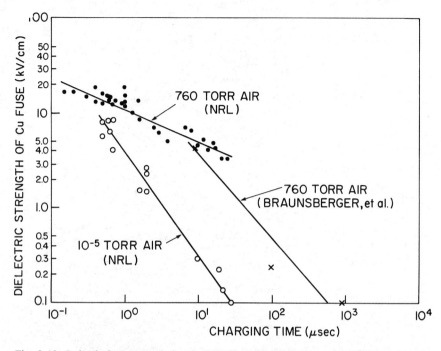

Fig. 8–13. Inductively generated electric field E, at restrike, shown as a function of t_e for fuses exploded in vacuum and at atmospheric pressure. Reproduced from Ref. 10.

to about 9 kV/cm at low pressure. This suggests that breakdown around the fuse rather than fuse expansion may be responsible for the low values of E_r in the Paschen regime ($10^{-3} \lesssim P \lesssim 10^1$ torr). The fuse data shown in Fig. 13 were obtained by exploding fuses in a low-pressure chamber (with no fuse 3 in the circuit). Fuse 3 was used, in turn, to measure E_b of the two types of gap electrodes. The half-width of the voltage pulse generated by either fuse 2 or fuse 3 was about 200 ns. The current and t_e associated with fuse 2 are 5–10 kA and 1.5 μs, respectively. Measurement of the electrode gap breakdown determined that, with the possible exception of the low breakdown region where data scatter is quite large, these electrodes do not affect the observed value of the restrike field. Hemispherical electrodes with low field enhancement were also tested to establish the maximum practical fields that could be supported if better fuse performance were obtained.

Figure 8–14 shows that the induced fields are about 9 and 20 kV/cm for fuses operating in a vacuum and in air respectively. Fuse recovery in both air and vacuum was studied in the range $0.6 \gtrsim \Delta t \gtrsim 10 \mu$s. In cases in which restrikes occur, both the current interruption (at 3.5 kA in air and

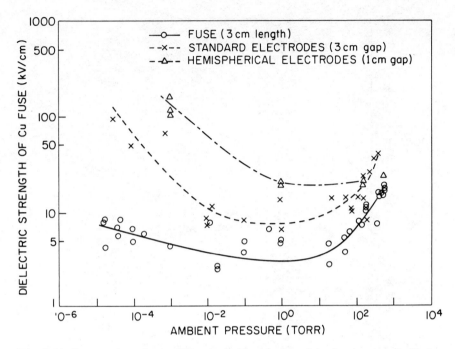

Fig. 8–14. Pressure dependence of E_r generated by opening of the fuse compared with the electrode gap breakdown field E_b. E_b was obtained for flat electrodes (standard) used in mounting of fuses. Also higher breakdown strength was obtained for rounded electrodes with lower field enhancement. Reproduced from Ref. 10.

2.1 kA in vacuum) and induced voltage waveforms have similar histories in air and vacuum. The recovery of the fuse in a vacuum, however, was found to be substantially different from that in air or water[5]. The difference appears to be due to the early onset of ionization, normally absent when media other than a vacuum are used.

The mechanism that determines the recovery field E_r of the fuse exploded in air involves the pressure and temperature distribution of the expanding fuse and the heated air that surrounds the fuse[28]. Therefore, E_r is expected to depend on the delay between vaporization (leading to current interruption and subsequent commutation to fuse 3, shown in Fig. 8–12) and reapplication of the voltage generated by fuse 3. This dependence is shown in Fig. 8–15. For small Δt, the recovery field E_r $(t_e + \Delta t)$ is approximately equal to the restrike field $E_r(t_e)$. At large Δt(11 μs), the recovery field is much smaller due to substantial expansion of the fuse channel. In fact, the dependence of E_r on Δt matches with the 1–2 kV/cm measured by Borisov

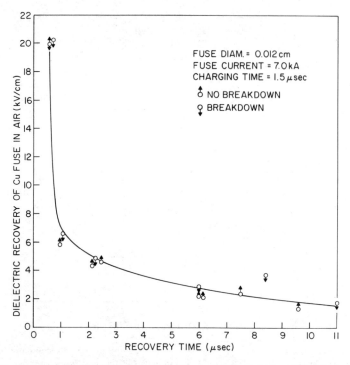

Fig. 8–15. Recovery field E_r of fuses exploded in air. The field is determined by converging the breakdown and no-breakdown data (by adjustment of fuse length) to a minimum differences. Reproduced from Ref. 10.

et al.[28] at Δt of about 60 μs for Cu fuses in air with a somewhat different history of charging current.

The characteristics of the recovery of fuses exploded in a vacuum are more complex, because from the time of the onset of vaporization there appears, in contrast to wires exploded in the air, strong ionization of the channel. Study of fuse behavior in a vacuum and in low pressure gases has been limited in the past, largely, to applications such as plasma, light, and x-ray source development[33], i.e., the purpose of these studies was other than the determination of current interruption properties. The most thorough studies of the electrical properties of fuses and the mechanisms responsible for them have been performed in Sweden[34, 35, 36]. The value of an inductively generated electric field was found to be about 5 kV/cm at an ambient gas pressure of 10^{-5} torr[36]. Reference 10 shows that the restrike field, E_r, for the Cu fuse decreases also with current pulse duration before the explosion of the fuse in a manner similar to that for wires exploded in atmospheric gases and in liquids[17], i.e., $E_r \propto t_e^{-\alpha}$ where $\sim 0.3 < \alpha < 1.0$ covers all cases. No studies of the dielectric strength recovery rate have been performed for fuses in low pressure environments.

Channel ionization could result from the thermionic emission of electrons and ions (and their acceleration in the potential drop across the fuse) during the heating and vaporization phase, as suggested in Refs. 35 and 36. The resistance of the vapor is lowered by ionization, relative to fuses in air. The time of onset of ionization is indicated in Ref. 34 and 36. The presence of ionization is also supported by quantitative resistivity measurements, during and after vaporization, which are consistent with the presence of free electrons[10]. Such behavior of the fuse in a vacuum makes it difficult to achieve total current interruption in the fuse. As current commutation is attempted, residual current continues through fuse 2.

To obtain the condition of total current cutoff, an additional short fuse (exploded in air) was added in series with fuse 2 (exploded in a vacuum). The air fuse provided a complete interruption of current in both series elements. This way, the voltage applied across the series fuses by the fuse 3 after delay time Δt provided a measure of the upper limit on the vacuum fuse resistance for Δt up to 10 μs. It was found that the resistance remains low and the recovery field is negligibly low. This conclusion is based on the observation that neither of these parameters differed from the values obtained when only fuses in air were used. This is also consistent with the recombination time for singly ionized plasma, where the three-body collisions of atomic ions and electrons are dominant. The recombination rate coefficient,[37] $R_r \simeq 8 \times 10^{-27} T_e^{-4.5}$, gives the scale decay time, $\tau = 1/R_r n_e^2$. For an expected density of the order of 10^{16} cm^{-3}, $\tau = 30$ μs, i.e., low channel resistance persists for times that are longer than available for testing.

In the experiments described here, the electrical characteristics of fuses exploded in air and vacuum were found to differ substantially; behavior in air is distinct from that in other media in which such effects as tamping of the fuse channel expansion influence the electrical properties. Fuse behavior in a vacuum differs more fundamentally from other cases because of the presence of ionization, nearly simultaneous with vaporization, which cannot be effectively inhibited. The description of the experiments indicates the flexibility available to the user of fuses as opening switches, as well as constraints imposed on such use.

Fuses Operated in Water.   It was recognized by Braunsberger et al.[17] that confinement of the fuse explosion by non-porous material with much higher density than gases should lead to higher breakdown potentials than a fuse in air at atmospheric pressure can support. Figure 8–16, based on the results from Ref. 17, is a logarithmic plot of the inductive field generated by fuses operating in air, water, and tamped water (i.e., the cylindrical water jacketing the fuse in turn confined by a strong wall) from left to right, respectively. The inductive fields, plotted as a function of time-to-explosion, t_e, are highest for shortest t_e. Extrapolation of the breakdown field values to a regime of very short t_e (10–100 ns) suggests that these values become independent of the medium surrounding the fuse. The magnitude of the extrapolated fields (including even that for the fuses exploded in vacuum as seen from Fig. 8–13) is in the 100–1000 kV/cm range and corresponds to breakdown field values associated with gases at high pressures. The reader is cautioned, however, that no published data exists to suggest that such fields can be achieved in the design of fuses for practical switching.

It was further recognized by the Naval Research Laboratory group[19] that in addition to high inductive fields, water confinement of fuses may also help in the development multi-stage opening switches which employ fuses for current interruption, as depicted, for example, by the schematic of Fig. 8–9. This group has made detailed measurements of the recovery of fuses

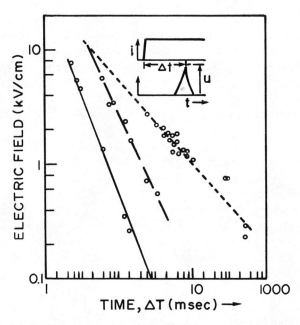

Fig. 8-16. Peak inductively generated electric field (in kV/cm) shown as function of the time to vaporization for fuses operating in air (left curve), in unconfined water (middle curve), and in water confined by a polyethelene tube (right curve). Reproduced from Ref. 17.

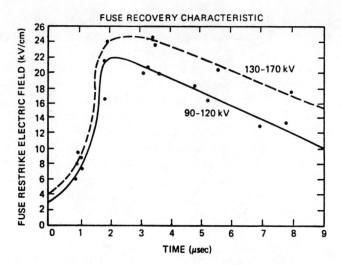

Fig. 8-17. Foil fuse high voltage recovery characteristics. The two curves correspond to commutation out of the foil at the 3 and 4 kV/cm self-stress range. The difference is due to imcomplete vaporization at lower fields and unnecessary enery dissipation at higher fields. Reproduced from Ref. 10.

exploded in water[5]. Figure 8-17 is a plot of the fuse recovery field (restrike level associated with the application of a voltage pulse after a delay relative to initially generated inductive fields). The field values were obtained for delay times of up to 9 μs. The curve in Fig. 8-17 is different from that obtained for fuses exploded in air and shown in Fig. 8-15; the recovery for fuses in water takes longer (~ 2 μs) to be established than for fuses in air. Once recovery has been achieved, it decays slowly in contrast to the rapid fall of the restrike field, as seen in Fig. 8-15. Use of these fuse recovery properties for a multi-stage opening switch for submegajoule inductive storage pulses is discussed in Ref. 5.

APPLICATION-ORIENTED PACKAGING

In contrast to closing switch technology, with which commercial products have been developed to supply a variety of switches needed in high power pulses, there are no standardized designs for off-the-shelf high power ($> 10^{10}$ W) fuse switches that go beyond the needs of the electric power distribution networks or specialized protection devices. Generally, when fuse switches are chosen for a given application, a customized design is employed, often requiring substantial development and testing.

Major applications for fuse switches in high power systems exploit the rapid opening of the fuse to optimize the impedance match between the energy storage system and the load. For example, fuses have been employed in the development of power sources for thermonuclear experiments to speed up the magnetic compression of plasma[3], in fast Z-pinches[4] and to develop techniques for current multiplication such as that developed by I. A. Ivanov et al.[7] in the USSR and by G. H. Tripoli et al.[38] in the U.S. Wire fuse switches can also be used for generation of intense relativistic electron beams. V. P. Kovalev et al.[2] and B. M. Kovalchuk et al.[43] have employed an inductive peaking of the capacitor bank (configured into a MV-level Marx circuit) to generate 2.5 MeV electron beam of about 30 kA for intense pulsed x-ray (*bremsstrahlung*) production. R. J. Blaher et al.[39] and Fell et al.[40] have employed fuse switches to study pulsed electron beam diode behavior at power levels in the 10^9–10^{10} W regime. To further emphasize the broad use of fuses, other recent applications are noted. One of these, relatively novel, is switching in circuits for production of high power ($> 10^{10}$ W) pulse bursts[41] with pulse-to-pulse separation of less then 10 μs. Another is that of T. Ito et al.[42] of Mitsubishi Electric Corporation of Japan, who invented a reusable fuse that operates as a current limiter in power lines (albeit at a much lower power level).

Table 8–1 lists examples of the major experimental facilities used for development of fuses for switching and their employment to generate single electromagnetic pulses. This listing is limited to facilities which use switches at power levels exceeding 10^{10} W. (Switching power level is defined as the product of the current being interrupted by the switch and the inductive voltage developed by the switch). With the exception of one, all the listed examples employ capacitors as current sources; the last example uses a homopolar generator as a very compact current source[44], providing a demonstration of an "inertial-inductive" storage system generating an output power usually associated with fast, large energy capacitor banks. The gen-

**Table 8–1. Examples of Application of Fuses
as Switches in Very High Power Pulse Systems**

Energy Storage		Fuse Current	Generated Voltage	Opening Time	Switched Power	Reference
1. Capacitors	60 kJ	260 kA	60 kV	500 ns	1.5×10^{10} W	8–25
2. Capacitors	400 kJ	240 kA	700 kV	100 ns	1.7×10^{11} W	8–5
3. Capacitors	1100 kJ	3000 kA	140 kV	500 ns	5.2×10^{11} W	8–3
4. Capacitors	9500 kJ	35000 kA	180 kV	500 ns	6.0×10^{12} W	8–4
5. Marx generator	130 kJ	60 kA	6000 kV	100 ns	3.6×10^{11} W	8–2
6. Homopolar	3000 kJ	70 kA	200 kV	2000 ns	1.4×10^{10} W	8–44

eration of the high power pulses, using either a capacitor or a homopolar current source, is possible through the use of fuse opening switches.

The realization of the high energy pulser systems listed in Table 8-1 (and many less well-documented systems using explosive generators as current sources) is based on a long development history of fuse packages which control the explosive forces generated by the fuses while at the same time striving to satisfy electrical performance constraints. Confinement of even a few or a few tens of kilojoules may become a problem when delicate diagnostic or experimental equipment is in the vicinity. Haarman et al.[25] have developed a disposable package for copper foil fuses which absorbs 20 kJ without rupture and could withstand 60 kV pulses. It uses a folded foil with 1.5 mm plastic between the folds to reduce switch inductance and two 1.2 cm layers of sand as an arc-inhibiting and shock-absorbing medium. Their test bed is sufficiently versatile to test other successful and unsuccessful ideas for packaging, as well as to correlate data between the packaging material and the resulting acoustic noise.

A similar approach, using shock-absorbing "quartz" sand, is in the development phase at the Air Force Weapons Laboratory for use with very large capacitor banks[4]. Results of the initial phase of this development are reported in Ref. 30. The package shown in Fig. 8-18, containing the fuse and the inductor, was operated at up to 450 kJ and hold-off 400 kV inductive pulses of short ($\sim$ 500 ns) duration. The fuse and storage inductor are packaged as shown in Fig. 8-18. The primary insulation for the output voltage is a 0.3 cm polycarbonate sheet overlayed with four layers of 0.012 cm polyester film (Mylar). The fuse material is 0.0025 cm aluminum foil whose width, constant throughout its length, is adjusted according to the scaling principles outlined in Ref. 15. The quenching material surrounding the hairpin folded fuse foil is quartz sand (glass beads), commercially available for "bead blasting" of nominally 60-100 μm diameter. The beads are contained in a folded polyester film envelope. Current connections to the

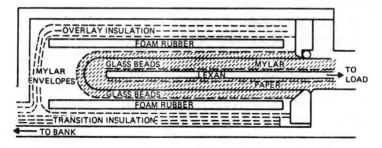

Fig. 8-18.   Fuse and storage inductor construction. Reproduced from Ref. 30. © 1981 IEEE.

ends of the foil are made by secure mechanical clamping. The entire fuse is assembled in place on the machine and confined by approximately 3 cm of moderate density polyurethane foam on each side to compress the package, absorb shock, and confine the beads. The entire assembly is situated between the output transmission lines of the system in the space which is part of the storage inductance.

The Air Force Weapons Laboratory group has extrapolated their fuse opening switch[4] to a 5 MJ level, using the initial low resistance of the fuse to charge a 10 nH inductor with peak current of 35 MA. Fuse packages interrupt the current in a few hundred nanoseconds, compressing substantially the output pulse time. The group's development work additionally explored the practicality of cooling foil fuses to as low as 80°K to enhance their performance. They found in this work[45] a significant effect on performance, arising from reduced fuse losses during the inductor charging time. Specifically, the effect of initial temperature on fuse performance occurs in the early-time resistivity of the fuse. Colder aluminum fuses open later than similar fuses at room-temperature, allowing the designer to use a narrower and presumedly a longer fuse in a cold environment and therefore achieve higher resistances as seen by the load. Heat transfer to the medium does not seem to have an effect on early time fuse performance. The various quenching materials investigated showed different late time effects, but this is a well known condition and is not unexpected[29].

Ivanov et al.[7] developed a fuse package integrated with a circuit breaker switch to serve as one of the modules for a 20 MJ system. The complete system had a lifetime of about 40 reliable switchings before the main components had to be replaced. The assembly was 90 cm long and approximately 30 cm in diameter.

Other types of switch packaging, handling smaller energies, were developed by Kotov et al.[1] for multi-MV applications, using fuses in nitrogen at up to 10 atm. To produce high peak power pulse bursts, Ford et al.[41] designed and tested packages for an array of fuses (and closing switch gaps) for sequential opening. Foil fuse assemblies used during switch development were functional but unwieldly. In order to demonstrate that such fuses could be produced simply, and would be easily replaceable, the assembly shown in Fig. 8–19 was constructed and tested. In this assembly, cemented lucite components contain multiple foil fuses in water medium, with each fuse separated by lucite partitions. A spark-gap closing switch electrode becomes an integral part of the foil clamp assembly. A six-section assembly, of 3.85 cm in width, was tested and shown to be both reliable and convenient to use. The output pulses generated by this switch were in the 50 kA, 100 kV range. Because the fuses were surrounded by low-conductivity water, their

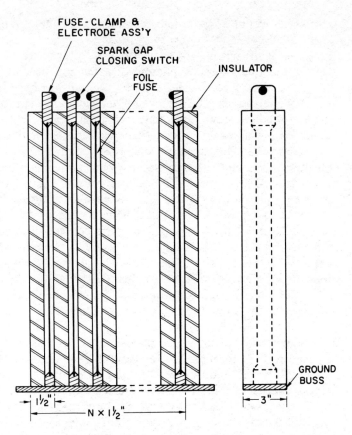

Fig. 8-19. Compact assembly for an array of fuses used in generation of high frequency pulse trains. Reproduced from Ref. 41. © 1979 IEEE.

operation could easily be extended to operating voltages in the megavolt regime.

PROGNOSIS

Exploding-wire research, of which fuse switching technology constitutes a very substantial part, has been active for a very long time. E. Nairne[46] performed in 1774 experiments with exploding wires to prove that the current in all parts of a series circuit is the same. Even though more than two centuries have passed from these initial attempts to understand the phenomena of electrically exploded conductors, there are a number of remaining issues, which, if resolved, would advance substantially the technology of fuse switching.

A potential for significant improvement of fuse performance has been revealed by investigation of the three phases associated with switching: conduction, non-linear resistance change, and recovery, all involving new as well as conventional materials. The description of conducting plastics at the end of Chapter 10 indicates one development that may lead to better fuse switches, involving plastics whose conductivity decreases dramatically as a result of molecular rearrangement in the material at around 120°C. There is no need to invest energy into vaporization to obtain switching using fuses made of such plastics. There is the likelihood that, because of low transition temperature, such "plastic fuses" would not explode violently, as do conventional fuses. Thus the techniques of packaging may become much simpler and, because of low switching temperature, there would be no ionization, hence possibly resulting in much higher restrike and recovery fields.

For example, better control of the exploding wire hydrodynamics during vaporization may provide higher restrike fields. Certainly, fuses which are more uniform along their length would provide faster switching times for given fuse configurations. Control of hydrodynamics through improved fuse fabrication techniques or other means could also provide much better recovery fields, which today do not exceed 20–30 kV/cm (for fuses in water[5]), apparently because of the formation of the Paschen minimum[47]. One valuable effort would be to extend the use of fuses in a vacuum to wider operating ranges. The issues that must be addressed to achieve this usage, while more complex, could lead to 100 kV/cm fuses (as suggested by extrapolation of the data for fuses in a vacuum given in Fig. 8–13). Such fuses would be competitive in terms of performance with erosion switches, discussed in Chapter 9.

Attempts to use fuses with a ratio of cold resistance to switching resistance of about 70, where cryogenically cooled stainless strips conduct for a time necessary for ohmic heating to bring them close to melting[48], show that the "re-usable" fuse is possible. Its practicality, in terms of energy expended in cooling and switching characteristics, is not proven and awaits development of other materials (such as special alloys). Such new materials might, for example, provide a high ratio of initial and final resistances and a high rate of resistance rise without explosive pressure.

Development of various hybrid switch designs can lead to substantial improvement in the performance of switching fuses. P. S. Levi et al.[32] employ foil fuses together with sheet explosives, as already mentioned, in such a way that explosive pressure confines the high pressure fuse vapor, preventing its expansion, with the result of a more than 3 fold increase in the inductively generated open circuit voltage. It is likely that by inducing effective turbulent mixing at the interface of vapor-explosive products, an even better rate of switch resistance rise could be achieved. Careful attention to the hydrodynamics of the explosive gases and the resultant flow of metal conductors (and their final rupture) suggests that another of the hybrid versions can be developed, where the explosive is used to stretch the metal, reducing the cross-section so that the current density can attain a level that vaporizes the remaining metal. This approach would combine the best features of explosively-driven switches, i.e., their very low resistance during conduction and low-loss conduction for long periods, with those of fuses that are capable of very fast opening.

Finally, it has been proposed by Bystrov et al.[50] that switches made by dispersing conducting metallic powder in the explosive matrix formed into sheets or cylinders could provide very fast (a few microseconds) interruption of megampere current. There appears to be no data available to support the design of switches based on this idea, however.

Better understanding of fuse physics and of techniques such as those used in the hybrids can be expected to lead ultimately to 1 MV/cm inductively generated fields, since both explosive products and cool high pressure vapor can maintain such fields without breakdown. Even though switch inductance would be reduced very considerably, diffusion of current in the conducting parts of the explosive and the finite rate of detonation would tend to limit the opening time to a few tens of nanoseconds.

REFERENCES

1. Yu. A. Kotov, B. M. Kovalchuk, N. G. Kolganov, G. A. Mesyats, V. S. Sedoi, and A. L. Ipatov, "High Current Nanosecond-pulse Electron Accelerator with Inductive Shaping Element," *Sov. Phys. Tech. Lett.* **3**, 359, (1977).
2. V. P. Kovalev, A. I. Kormilitsyn, A. V. Luchinskii, V. I. Martynov, and A. I. Pekhterev, "IGUR-1—An Electron Accelerator Using Inductive Energy Storage and Exploding Wires," *Sov. Phys. Tech. Phys.* **26**, 1082 (1981).
3. V. A. Burtsev, A. B. Berezin, A. P. Zhukov, V. A. Kubasov, B. V. Lyublin, V. N. Litunovskij, V. A. Ovsiannikov, A. N. Popytayev, A. G. Smirnov, V. G. Smirnov, V. P. Fedyakov, and V. P. Fedyakova, "Heating a Dense Theta-Pinch Plasma with Strong Fast Rising Magnetic Fields," *Nucl. Fusion* **17**, 887 (1977).
4. R. E. Reinovsky, W. L. Baker, Y. G. Chen, J. Holmes, and E. A. Lopez, "SHIVA Star Inductive Pulse Compression System," *Digest of Technical Papers of the Fourth IEEE Pulsed Power Conference,* IEEE Cat. No. 83CH 1908-3, p. 196 (1983).
5. D. Conte, R. D. Ford, W. H. Lupton, and I. M. Vitkovitsky, "TRIDENT—A Megavolt Pulse Generator Using Inductive Energy Storage," *Digest of Technical Papers of the Second IEEE International Pulse Power Conference,* IEEE Cat. No. 79CH 1505-7, p. 276 (1979).
6. R. D. Ford, D. J. Jenkins, W. H. Lupton, and I. M. Vitkovitsky, "Multi-Megajoule Inductive Storage for Particle Beam Production and Plasma Implosions," *High Power Beams 81,* H. J. Doucet and J. M. Buzzi, eds., in *Proceedings of the Fourth International Con-*

ference on High Power Electron Beams Research and Technology, Palaiseau, France, p. 743 (1981).

7. I. A. Ivanov, A. P. Lotoskii, N. P. Pugachev, and V. A. Trukhin, "High Power Three-Stage Switch for an Electrical Discharger with Inductive Energy-Saving Element," *Instrum. and Exp. Techniques* **25** (1), 895–897 (1982). (Same as Ref. 14, Ch. 2)

8. T. H. Lee, *Physics and Engineering of High Power Devices,* MIT Press, Cambridge, MA (1975).

9. K. F. McDonald, T. Smith, J. Golden, and T. Conley, "Measurements of Fuse and Resistor Characteristics for Multi-Megajoule Capacitor Bank Application," *IEEE Plasma Science Abstracts,* Abstract 3B-7, (1986). The fuse package specifications are given in the McGraw-Edison Company (Olean, NY) Power Systems Division Apparatus Catalog 240-60, p. 2 (March 1978).

10. I. M. Vitkovitsky and V. E. Scherrer, "Recovery Characteristics of Exploding Wire Fuses in Air and Vacuum," *J. Appl. Phys.* **52**(4), 3012–3015 (1981).

11. Yu. A. Kotov, N. A. Kolganov, V. S. Sedoi, B. M. Kovaltchuk, and G. A. Mesyats, "Nanoseconds Pulse Generators with Inductive Storage," *Proceedings of the First IEEE International Pulsed Power Conference,* Lubbock, TX, IEEE Cat. No. 76H1147-8 REG-5 (1976), and Yu. A. Kotov, B. M. Kovalchuk, N. G. Kolganov, G. A. Mesyats, V. S. Sedoi, and A. L. Ipatov, *Sov. Tech. Phys. Lett.* **3**, 359 (1977).

12. J. Benford, H. Calvin, I. Smith, H. Aslin, "High Power Pulse Generation Using Exploding Fuses," *Energy Storage Compression and Switching,* W. H. Bostick, V. Nardi, and D. S. F. Zucker, eds., pp. 39–49, Plenum, New York (1976).

13. I. M. Vitkovitsky, D. Conte, R. D. Ford, and W. H. Lupton, "Current Interruption in Inductive Storage Systems Using Inertial Current Source," *Energy Storage Compressing and Switching,* Vol. 2, V. Nardi, H. Sahlin, and W. H. Bostick, eds., Plenum, New York, pp. 953–971, (1983).

14. I. M. Vitkovitsky, D. Conte, R. D. Ford, and W. H. Lupton, "Inductive Storage for High Power REB Accelerators," *Proceedings of the Second International Topical Conference on High Power Electron and Ion Beam Technology,* Vol. 2, Cornell Univ. Press, Ithaca, NY, pp. 857–867 (1977).

15. C. Maissonier, J. G. Linhart, G. Gourlan, "Rapid Transfer of Magnetic Energy by Means of Exploding Foils," *Rev. Sci. Inst.* **37**, 1380 (1966).

16. W. H. Lupton, R. D. Ford, and I. M. Vitkovitsky, Plasma Physics Division, Naval Research Laboratory, Unpublished note (1983).

17. U. Braunsberger, J. Salge, and U. Schwarz, "Circuit Breaker for Power Amplification in Poloidal Field Circuits," *Proceedings of the Eighth Symposium on Fusion Technology,* Jutphaas, Netherlands, published by the Commission of the European Communities, Luxembourg, pp. 399–406 (1974).

18. W. H. Lupton, Internal notes and working papers, Naval Research Laboratory (1980).

19. D. Conte, R. D. Ford, W. H. Lupton, and I. M. Vitkovitsky, "Two-Stage Opening Switch Techniques for Generation of High Inductive Voltages," *Proceedings of the Seventh Symposium on Engineering Problems of Fusion Research,* Vol. II, IEEE Pub. 77CH1267-4-NPS, pp. 1066–1070 (1977).

20. L. Vermij, "The Voltage Across a Fuse During the Current Interruption Process," *IEEE Trans. Plasma Sc.,* **PS-8**, 460–468 (1980).

21. J. D. Logan, R. S. Lee, R. C. Weingart, and K. S. Yee, "Calculation of Heating and Burst Phenomena in Electrically Exploded Foils," *J. Appl. Phys.* **48**, 621–628 (1977).

22. R. D. Ford, D. Jenkins, W. H. Lupton, and I. M. Vitkovitsky, "Operation of Multi-Megajoule Inertial-Inductive Pulser," *Digest of Technical Papers of the Third International Pulsed Power Conference,* Albuquerque, NM, T. H. Martin, and A. H. Guenther, eds., IEEE Cat. No. 81CH1662-6, pp. 116–121 (1981).

23. The author is grateful to W. H. Lupton and R. D. Ford of the Naval Research Laboratory for making available their data on fuse physics and its development as part of the work in support of the Advanced Technology Project, Defense Nuclear Agency (1984).

24. Yu. A. Kotov, N. G. Kolganov, V. S. Sedoi, B. M. Kovalchuk, and G. A. Mesyats, "Nanosecond Pulse Generators with Inductive Storage," *Proceedings of the IEEE International Pulse Power Conference,* Lubbock, TX, IEEE Cat. No. 76-CH1147-Reg-5 (1976).

25. R. A. Haarman, R. S. Dike, and M. J. Hollen, "Exploding Foil Development for Inductive Energy Circuit," *Proceedings of the Fifth Symposium on Engineering Problems of Fusion Research,* Princeton University, pp. 459–461, IEEE Publ. No. 73 CH0843-3-NPS (1973).

26. V. A. Burtsev, V. N. Litunovskii, and V. F. Prokopenko, "Electrical Explosion of Foils I," and "Electrical Explosion of Foils II," *Sov. Phys. Tech. Phys.* **22,** 950–956 and 957–961 respectively (1977).

27. J. R. Greig, R. E. Pechacek, M. Raleigh, I. M. Vitkovitsky, R. F. Fernsler, and J. Halle, "Interaction of Laser-Induced Ionization with Electric Fields," *Proceedings of the AIAA Thirteenth Fluid and Plasma Dynamics Conference,* Snowmass, CO, AIAA-80-1380 (1980).

28. R. K. Borisov, V. L. Budovich, and I. P. Kushekhin, "Restoration of the Dielectric Strength Following Explosion of a Wire," *Sov. Tech. Phys. Lett.* **3,** 516 (1977).

29. R. P. Henderson, D. L. Smith, and R. E. Reinovski, "Preliminary Inductive Energy Transfer Experiments," *Digest of Technical Papers of the Second IEEE International Pulse Power Conference,* IEEE Cat. No. 74 CH1505-7, p. 347 (1979).

30. C. Stuerke, R. E. Henderson, D. L. Smith, and R. E. Reinovski, "Multi-Megamp, Multi-Terawatt Inductive Pulse Compression System Operation," *Digest of Technical Papers of the Third IEEE Pulse Power Conference,* IEEE Cat. No. 81 CH1662-6, p. 328 (1981).

31. L. Niemeyer, "Experimental Studies on the Discharge Propagation of Exploding Wires in a High Vacuum and Determination of the Developing Plasma Properties," Doctoral Thesis, Technische Hochschule, Hannover (1969), published as an English translation in NASA Report TT F-13, 346, Washington, DC (1970).

32. P. S. Levi, J. D. Watson, M. L. Hines, and R. E. Reinovsky, "Staged Explosively Driven Opening Switch Development for Explosive Flux Compression Generators," *Digest of Technical Papers of the Fourth IEEE Pulsed Power Conference,* Albuquerque, NM, M. F. Rose and T. H. Martin, eds., IEEE Cat. No. 83CH1908-3, pp. 782–785 (1983).

33. *Exploding Wires,* W. G. Chase and K. Moore, eds., Vols. 1–4 (Plenum, New York, 1968).

34. B. Stenerhag, S. K. Handle, and I. Holmstrom, "Exploding Wires in Air and Vacuum," *Z. Phys.* **198,** 172 (1967).

35. I. Holmstrom, S. K. Handle, and B. Stenerhag, "Undegassed and Degassed Exploding Tungsten Wires in Vacuum," *J. Appl. Phys.* **39,** 2998 (1968).

36. B. Stenerhag, S. K. Handle, and B. Gohle, "Effect of Wire Parameters on the Emission of Hard X-Rays from Exploding Wires," *J. Appl. Phys.* **42,** 1876 (1971).

37. Y. B. Zeldovich and Yu. P. Raizer, *Physics of Shock Waves,* Vol. I., Academic Press, New York (1966).

38. G. H. Tripoli, P. J. Turchi, D. Conte, and S. W. Seiler, "Design and Operation of 0.4 MJ Current Multiplier Inductive Store with Multiple Exploding Foil Switches," *Digest of Technical Papers of the Fifth IEEE Pulsed Power Conference,* IEEE Cat. No. 85C2121-2, pp. 22–25 (1985).

39. R. J. Blaher and R. E. Reinovsky, "Inductive Store/Transformer Driven Diode System," *Digest of Technical Papers of the Fourth IEEE Pulsed Power Conference,* IEEE Cat. No. 83CH1908-3, pp. 126–129 (1983).

40. B. Fell, R. J. Commisso, V. E. Scherrer, and I. Vitkovitsky, "Repetitive Operation of an Inductively-Driven Electron Beam Diode," *J. Appl. Phys.* **53,** pp. 2818–2824 (1982).

41. R. D. Ford and I. M. Vitkovitsky, "Inductive Storage Pulse-Train Generator," *IEEE Trans. on Electron Devices* **ED-26**, 1527–1531 (1979).
42. T. Ito, T. Miyamoto, Y. Wada, Conference Paper No. C-72 103-5, presented at IEEE Switchgear Committee of the IEEE Power Engineering Society Winter Meeting, New York, (1972).
43. B. M. Kovalchuck, Yu. A. Kotov, and G. A. Mesyats, "High Power Nanosecond Electron Accelerator with Inductive Storage," *Sov. Phys. Tech. Phys.* **19**, 136–137 (1974).
44. R. D. Ford, D. Jenkins, W. H. Lupton, and I. M. Vitkovitsky, "Pulsed High-Voltage and High-Current Outputs from Homopolar Energy Storage System," *Rev. Sci. Inst.* **52**, 694–697 (1981).
45. J. C. Bueck and R. E. Reinovsky, "The Effects of Cryogenic Initial Temperatures on Aluminum and Copper Electrically Exploded Foil Fuses," *Digest of Technical Papers of the Fourth IEEE Pulsed Power Conference,* IEEE Cat. No. 83CH1908-3, p. 122 (1983).
46. E. Nairne, *Phil. Trans. Roy. Soc.* (London) **64**, 79–89 (1774).
47. I. M. Vitkovitsky, "Fuses and Repetitive Current Interruption," *Proceedings of the Workshop on Repetitive Opening Switches,* Tamarron, CO, M. Kristiansen and K. Schoenbach, eds., Texas Tech University, Lubbock, TX, pp. 269–283, (1981).
48. M. I. Bystrov, B. A. Larionov, V. P. Sinin, F. M. Sperakova, and A. M. Stolov, "Pulsed Power Sources Based on Transformer Inductive Energy Storage Devices with Non-Linear Elements," No. 2 in the Series *Inductive Energy Storage Devices and Switching Apparatus for Thermonuclear Installation,* Report of a Joint USSR-USA Seminar, USSR Atomic Energy State Committee, NIIEFA, Leningrad (1974). Translated by W. J. Grimes, P. O. Box 55, Hingham, Mass (1975). Paper 5 in this series also deals with use of "Non-Exploding Fuses" as switches.
49. A. S. Gilmour and J. D. Marshal, "Liquid Nitrogen Cooled Wires as Switchable High Power Direct Current Limiting Elements," *Proceedings First IEEE International Pulsed Power Conference,* IEEE Cat. No. 76CH1147-8 Reg 5 (1976).
50. M. N. Bystrov, V. A. Krylov, B. A. Larionov, V. P. Silin, and A. M. Stolov, "Impulse Power Sources for Feeding the 'Pinch with Liner' Installation," Report II, *Joint Soviet-American Seminar on Pulsed Fusion Reactors,* ESRII, Leningrad (1975).

9
Opening Plasma Switches

RATIONALE

This chapter reviews the research on developing opening reusable switches with performance characteristics that substantially differ from those with multi-shot capability, discussed in Chapter 7. Performance characteristics which recently have stimulated the development of plasma opening switches are the risetime and very high power capability promised by approaches involving changes in the state of plasma. Potential use of plasma changes for the interruption of current has been explored in response to the realization that volume discharge, in contrast to channel discharge such as that represented by arcs in conventional circuit breakers, provides a means of reducing the *density* of energy deposited into the conducting medium during the opening of the switch. Reduced deposition lowers heating and ionization of the medium and maintains low current density at the electrodes, so that new material is prevented from injection into the switching gap at unacceptable rates.

There are two distinct types of switches discussed in this chapter that fall in the categories of volume discharge. The first of these is a switch employing gas filled gaps with ionization provided by an externally injected low current density electron beam.[1] The ionized gas leads to conduction of current from an external circuit if sufficient voltage is maintained across the switch to overcome the ohmic drop associated with the finite conductivity of the partially ionized gas. Removal of the ionizing agent leads, under proper conditions, to a rapid increase in the resistivity of the gas even with a high (but less than breakdown) value of electric field remaining across the gas. This type of transition in plasma conductivity can, of course, be used to operate the switch as a closing switch. As the following discussion indicates, it is also possible to use electron beam ionization to operate a repetitively opening or closing switch.

A second type of switch uses injection of plasma into the interelectrode region to initiate conduction[2,3]. Such switches rely on very high currents to terminate conduction by a mechanism known as "erosion." The plasma opening erosion switch (PEOS), which can be operated only for very short (<1 μs) conduction time is complemented by a third type of switch, called the *plasma flow* or *plasmadynamic switch,* with a long conduction time that takes advantage of the strong force, generated by the magnetic field associated with the current being switched, to move the plasma away from the electrodes[4]. This is not a volume discharge switch. The high currents and long conduction time of the plasmadynamic switch require rather high plasma density; such plasma is provided by initially starting conduction through a thin metal foil which subsequently vaporizes and accelerates.

The switches discussed here are applied in high power inductive storage systems, where an inductor can be charged in a short time or where some other switch provides the necessary conduction time for charging. Examples of the three types of switches are discussed later. The operating range of the plasma opening switches when they are fully developed will extend from 10–100 kA for electron beam controlled switches (EBCS)[5,6] which can hold-off several hundred kilovolts and open in times shorter than 100 ns, to 10–20 MA and hundreds-of-nanoseconds opening times for the plasma-dynamic switch[4]. The least known characteristic of the latter switch is its voltage characteristic. It is not yet clear how large an inductive voltage can be generated by interruption of current in an inductive system or what recovery level can be achieved in times relevant for a given application. Finally, the plasma erosion opening switch (PEOS) has already operated in an intermediate current regime (~ 1 MA range). Used as a current steepening switch, it can open up in ≤ 10 ns. In such applications, it has withstood voltages up to 8 MV. In switch-opening applications, PEOS has routinely operated at a 3 MV level. These two cases are discussed in Ref. 7.

The plasma switches viewed as circuit elements serving to transfer energy from storage (in an inductor) to a load can be characterized by the energy dissipated in the switch (during conduction and during opening) and as a power amplifying element (i.e., as an element capable of increasing the voltage across the load relative to that initially across the storage element— with a nominal reduction in the current). Characterization in those terms requires, in addition to a consideration of switch mechanisms (i.e., transition from low to high resistivity), that the relationship between circuit elements and the switch be understood. Therefore, the mechanisms of switching and of parameters associated with the mechanisms and the role of the circuit is analyzed here in considerable depth.

The developmental nature of plasma switches at the present time suggests an emphasis here on scaling relations and broad coverage of the data, so that readers can evaluate the practicality of the switch designs for their applications. Various examples and hints, from the numerous references listed here, are provided as a guide for such designs. Additionally, some aspects of the switching performance of other types of switches are compared with the plasma switches, specifically with EBCS, to provide the reader with other options.

ELECTRON-BEAM CONTROLLED SWITCHES

In this section, the principles of operation of the e-beam controlled switch are discussed and then applied to some examples. The discussion follows closely that in Ref. 6, developed chiefly by one of its authors, R. J. Commisso. Special attention is paid to its capability for rapid recovery—the essential requirement for repetitively pulsed systems, whether inductively or capacitively driven. Detailed explanations of the switch physics can be found in several references[8, 9, 10, 11]. The e-beam controlled switch is compared to other candidate switches for repetitive, high power application. The design requirements for two applications of this switch to devices with characteristics similar to either the Experimental Test Accelerator (ETA) or Advanced Test Accelerator (ATA)[12] are then outlined. Next, a formalism for switch design that combines circuit requirements with switch physics is developed. With the use of this formalism, which is backed by a recently acquired data base[8, 9, 10, 11], the necessary e-beam controlled switches are "designed", i.e., the electrode geometry, gas pressure, and gas type to perform a given job are obtained.

The reader is cautioned that the design examples chosen do not necessarily make optimum use of the potential capabilities of the e-beam controlled switch concept. A full system study in which the e-beam controlled switch is included as an integral part of the design process from inception would be required for such a task[6]. Rather, the examples chosen serve to illustrate the practical engineering aspects of the switches, define some of the technical requirements necessary for their implementation, and indicate the steps that are necessary for a complete system design.

The results of this study indicate that e-beam controlled gas induction can be readily designed for a chosen demonstration of repetitive (> 10 kHz), high power (200 kV, 20 kA) switching. Further research and development concerning gas chemistry and atomic physics, cumulative heating in the switch, and the e-beam injector for the switch under repetitive, long conduction time (with respect to the load pulse width) conditions may be needed, depending upon the specific switch application.

Principles of Operation

Figure 9–1 shows schematically the components of an EBCS in a circuit consisting of a current source that charges an inductor when the switch is

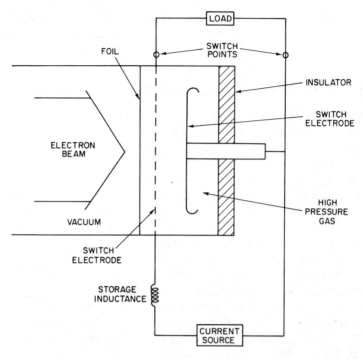

Fig. 9-1. Schematic of the electron beam controlled switch showing the main components. An externally generated beam is injected into a gas chamber filled with a gas separating the switch electrodes. One of the electrodes is perforated to allow the electrons to enter the inter-electrode region and ionize the gas. The switch current initially charges the storage inductor, and upon opening transfers the current to the load connected across "switch points." Reproduced from Ref. 5.

closed, i.e., it conducts as a result of ionization by an electron beam injected into a chamber containing a mixture of an attaching and a non-attaching gas. The ionization of the gas produced by the e-beam pulse competes with the attachment and recombination processes controlling the conductivity of the gas. The conductivity and hence the switch current is turned on and off in association with the e-beam pulse. In the case shown in Fig. 9-1, the opening of the switch transfers the current to a load.

An important distinguishing feature of a switch based on this concept is its ability to open (cease conduction) under high applied voltage. To avoid avalanche ionization, the switch must be designed in such a way that the maximum expected voltage across the switch is below the static self-sparking threshold (for transient voltages of less than 1 μs duration, the threshold is significantly higher, permitting this requirement to be relaxed).

As the discharge evolves, cumulative gas heating also must be constrained so that thermal ionization and, more importantly, local hydrodynamic reduction in gas density do not significantly lower the self-sparking threshold. A self-sustained discharge is thus prevented. Under these conditions, the fractional gas ionization, and thus switch resistivity, at any time is determined by the competition between ionization provided by the beam and the various recombination and attaching processes characteristic of the specific gas mixture, pressure, and applied electric field.

A second important feature of this switch is the volume discharge property. This characteristic makes it possible to avoid excessive heating of electrodes and the switch gas (as well as lessen mechanical shock and minimize the switch inductance). All these features combine to permit the discharge to return to its initial state of high resistivity very quickly once the source of ionization is removed. Unlike an arc disharge, this transition can be accomplished rapidly under an applied voltage. Some details of the gas chemistry and atomic physics associated with the switch operation and their coupling to the switch circuitry can be found in Ref. 9.

From a different perspective, the e-beam controlled switch behaves as a current amplifier. That is, the small (~ 1 kA) electron beam current can control a large (~ 10 kA) switch plasma current. The ratio of these two currents is called the *current gain, ϵ*. For $\epsilon \gg 1$, a substantial energy gain (energy delivered to the load/energy dissipated) can be achieved, as discussed below.

Comparison to Other Switches

A summary of high power closing and opening switches with potential application to repetitively pulsed systems is provided by Burkes et al.[13] There are basically six switch types which emerge as candidates for a high power repetitively pulsed switch: (1) the low pressure gas switch,[14] (2) the surface flashover switch,[15] (3) the thyratron,[13] (4) the high pressure spark gap discussed in Chapter 3, (5) the magnetic switch[16,17], and (6) the EBCS. The PEOS, although not yet tested, is also a likely candidate for repetive operation[13].

The ongoing research for both the low pressure gas and surface flashover closing switches has yielded some encouraging results. The technology appears to be simple[15]. At present, however, recovery times of only $\geq 100 \ \mu s$ have been observed for both devices with no applied voltage. Jitter may also be a problem. Under repetitive burst operation, both switches may have to be cooled. For the surface flashover switch, the insulator may additionally require cooling and the insulator lifetime is likely to be limited.

The thyratron is a well developed device; however, the present limitations in voltage, current, and risetime probably make it unsuitable for implementation in very high power systems. Further, the power consumption and long warm up period for the cathode heater may be disadvantageous.

The high pressure spark gap is the traditional choice for a closing switch for pulsed high power applications. It cannot be used as an opening switch. It can be very simple mechanically and has associated with it an extensive data base. The problem areas are the substantial gas flow requirements and energy dissipation associated with this device when used in high frequency repetitive high power applications. Also, a trigger pulse amplitude comparable to the applied switch voltage is necessary for low jitter.

Magnetic switches are now a part of the ATA design[16]. The potential advantages of such a switch are its simplicity, ruggedness, and lifetime. Questions still remain concerning overall size and weight, evaluation of core materials, limiting output pulse duration, pulse compression ratios, trade-offs in number of stages in cascade, and core bias and/or resetting methods.

An understanding of the physics of the EBCS along with the results of experiments leads one to conclude that the principal advantages of the e-beam controlled switch are: (1) rapid recovery (opening) of the switch when the electron beam ceases, and the consequent opening and closing repetitive capability[9,10,11] at high repetition rates; (2) negligible switch jitter[5,18]; (3) volume discharge resulting in low inductance and reduced switch current density (limiting electrode wear, switch gas heating, and mechanical shock); and (4) high power switching capability (for limited pulse bursts)[5].

Potential problem areas for this switch concept include the switch e-beam driver complexity, the effect of cumulative heating on the switch's repetitive capability, and switch packaging. The e-beam injector has modest peak power requirements on a per-pulse basis (200 kV, < 1 kA, $< 10^8$ W); however, it must provide a beam with the pulse shape and repetition rate required for the desired switching characteristics. Thermionic e-beam generators, developed for excitation of high power lasers, now provide[19,20] 1–5 μs e-beam pulses with ~ 100 ns rise and decay times at a 25 Hz repetition and at 10 A/cm^2. The repetition rate can probably be improved, e.g., if a lower e-beam current density is desired. With some further development, these devices may indeed provide the desired beam modulation characteristics. Thin film field emission cathodes with molybdenum cones[21] appear attractive because of their low control voltage (100–300 V), high current density (~ 10 A/cm^2), and continuous operation capability. An inductively driven electron beam system has successfully demonstrated two-pulse "burst" operation, generating two ~ 1 kA pulses with a 150–200 μs (limited by diode closure) interpulse separation.[22]

The effects of cumulative heating in the switch gas are not well known. Preliminary experiments[11,23], indicate that at least two pulse operation is possible at a deposited energy density of ~ 0.1 J/cm^3. These experiments also show a favorable (approximately linear) scaling of the maximum energy deposited before switch breakdown occurs, with increasing switch ambient pressure.

Although a stand-alone switch packaging scheme has been outlined[24], a complete system design has not yet been addressed. Such a design would be heavily dependent on the specific application and would have to incorporate the switch as an integral part of the system ab initio. At present there appears no fundamental limitation resulting from packaging considerations that would prevent integration of this switch into pulse power systems. For example, a compact, high pressure, e-beam controlled laser system with an e-beam generator (single pulse) compatible with severe optical requirements has been successfully assembled and operated[25].

Although much work is still needed for all of these switch concepts, the EBCS, in comparison with the other switch candidates, suffers from no fundamental flaw. In fact, in some areas (e.g., recovery) it has some very substantial advantages. It is especially promising in applications which require high repetition rate (>10 kHz) opening switches for short burst operation or very fast (<30 ns) opening times.

Design Criteria

In this section, performance characteristics, upon which a switch can be designed, are outlined. These characteristics strongly depend upon the specific application of the switch. Optimum switch performance requires that the switch be incorporated into a system design at its inception. However, to illustrate the practical design considerations for this switch concept, the requirements of the ETA/ATA inductive accelerator or beam generator[12] are chosen as examples. In this context, the requirements for application of this switch are described for three energy storage schemes: capacitive, hybrid, and inductive, each defined in forthcoming figures. Depending upon the specific requirements, each scheme may involve, respectively, a more stringent set of switch performance and e-beam injector characteristics and a progressively greater extrapolation from the present data and technology base.

Capacitive Systems—Application of the Repetitive Closing Switch Mode.   The e-beam controlled switch (EBCS) based on the principles outlined above can be used at high power levels ($\sim 10^{10}$ W) with capacitive energy storage as a closing switch, in applications where very fast risetime pulses of short duration (<100 ns) must be generated at high repetition frequency (>10 kHz).

One example of such an application is an output switch of the ETA/ATA. For definition purposes, the load characteristics are assumed to be those summarized in Table 9-1[12]. The characteristics displayed in this table imply that the switch e-beam injector has sufficiently fast risetime and that the working gas can ionize sufficiently fast in response to the injected beam.

Two circuits which provide the required output for the load are shown in Fig. 9-2. Both circuits depend on use of switch S_1 for charging. Typically, direct Marx charging as given in Fig. 9-2a can provide a charge time for capacitor C_0, of $\tau_{CH} = \sim 1$ μs. Using the voltage step-up transformer (Fig. 9-2b) has no deleterious effect other than to possibly increase τ_{CH} to several microseconds. Electrically, the switch must deliver a peak power $P = 4 \times 10^9$ W to the load for each pulse. With a > 10 kHz repetition rate (< 100 μs pulse-to-pulse separation), the average power output of the pulse is $P_{av} = P (\tau_p/\tau_{pp}) \simeq 1 \times 10^6$ W. The power transport through the switch leads to some dissipation of energy in the switch. The amount that is dissipated depends on the pulse duration, the current, and the switch voltage drop during conduction, and is less than the energy delivered to the load. The opened state of the switch has negligible conduction in all cases considered here. Additionally, the switch inductance must be limited to $L_{SW} \ll V_L$ $\tau_R/I_L \simeq 100$ nH, as suggested by the parameters in Table 9-1. A schematic representation of the time history of the switch resistance, R_{SW}, is shown in Fig. 9-3.

Finally, to obtain the necessary time variation of the gas conductivity of the switch, the electron beam injected into the gas must satisfy the pulse shape, repetition rate, and current requirements. The e-beam current is set by a number of factors, as discussed below under "Switch Designs." The beam modulation was discussed under "Comparison to Other Switches" above. It is, of course, assumed for the remainder of the discussion that an appropriate e-beam source is available for controlling the switch.

**Table 9-1. Summary of ETA/ATA Module
Load Characteristics.
(Closing Switch, Capacitive System)**

Type of load	Electron beam diode
Load voltage, V_L	200 kV
Load current, I_L	20 kA
Equivalent impedance, R_L	10 Ω
Pulse duration, τ_p	40 ns
Pulse risetime, τ_R	10 ns
Pulse-to-pulse time, τ_{pp}	< 100 μs
Pulses per burst, n	≥ 5

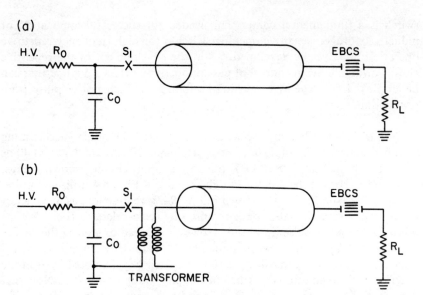

Fig. 9-2. Circuit schematics illustrating the application of EBCS in the closing mode. The energy for the output pulses is initially stored in a capacitor and transfered to the intermediate pulse-forming line (a) directly and (b) via a step-up transformer. The pulse-forming line output is controlled by EBCS.

Inductive Systems—Application of the Repetitive Opening Switch Mode.

In projecting such applications as those of charged particle beam technology to practical systems, the size and weight of the system are major considerations to the designer. The power source represents a major component of any of the proposed systems. Because inductive storage offers a potential for much more compact designs than those that would be possible with capacitive systems as discussed in Chapter 2, there exists a strong incentive to develop the necessary inductive storage components. An opening

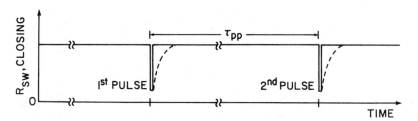

Fig. 9-3. Illustration of the periodic drop of switch resistance (in response to the repetitive injection of ionizing electron beam) for producing a repetitive output in Figs. 9-2a and 9-2b.

switch is a fundamental component needed for successful application of inductive storage; repetitive opening switches for high frequency repetitive operation must yet be developed. As in the development of the repetitive closing switch, e-beam control of gas conductivity offers a method that can be employed for repetitive opening switching for high power pulse train production.

Hybrid Pulser. For an accelerator consisting of modular accelerating sections, described in Table 9–1, each producing 20 kA, 200 kV, and 40 ns pulses at a burst rate of > 10 kHz, the power supply circuits, shown in Figs. 9–2a and 9–2b, can be employed. The circuit in Fig. 9–4 utilizes a pulse-forming line (similar to that shown in Fig. 9–2), represented as a capacitor C, to form a specific pulse shape required by the accelerator (i.e., risetime of 10 ns and 40 ns pulse duration). However, instead of storing the energy initially in a capacitor, now the inductor, L_0, is initially charged by a current, I_0, through the explosively actuated switch[26, 27], (denoted by E in the diagram, similar to those described in Chapter 7) for a time τ^0_{CH}. Note that some current generators, e.g., a homopolar[28], may require earlier, additional stages of pulse compression. When switch E opens, the current is commutated (in a time τ_{COM}) to the next stage which contains the EBCS(O) operating in the repetitive opening mode. EBCS(O) is closed and conducting during τ_{COM}. After the commutation time, EBCS(O) is commanded to open. The opening generates an inductive voltage, and the pulse line C is charged for the time τ_{CH}. At the end of τ_{CH}, EBCS(O) again closes.

The output for the pulse line is an e-beam controlled device already described in the previous section, i.e., it operates in the repetitive closing mode (denoted by the letter C in paranthesis). Fig. 9–5 is a schematic representation of the time histories of the resistance of EBCS(O) and EBCS(C) of Fig. 9–4. Thus, the circuit in Fig. 9–4 employs two repetitive EBCS's. The EBCS(O) has a long conduction time, equal to $\tau_C = (\tau_{pp} - \tau_{CH})$, and a

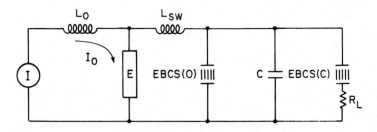

Fig. 9–4. Circuit diagram illustrating the application of an EBCS in opening mode EBCS(O) and in closing mode, EBCS(C) for a hybrid inductive-capacitive storage system.

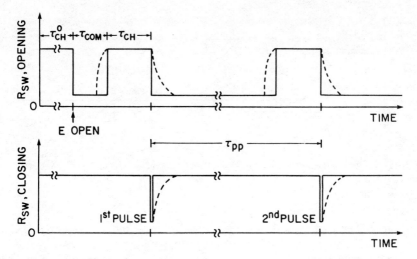

Fig. 9-5. Time behavior of the resistance of the switching EBCS(O) and EBCS(C), top and bottom diagrams respectively, as employed in the circuit of Fig. 9-4.

short non-conduction time, $\tau_{NC} = \tau_{CH}$, during which the capacitor C is charged. The EBCS(C) is identical in its operation to the switch discussed in the section "Capacitive Systems—Application of Repetitive Closing Switch Mode."

Inductive Pulser. A more compact and simpler pulser would result if the circuit in Fig. 9-6 could be used. The pulser represented by this circuit is purely inductive, eliminating any need for capacitive storage. However, this scheme represents the farthest extrapolation of the present data and technology base thus far considered. Figure 9-7 is a schematic representation of the time history of the opening switch resistance for a purely in-

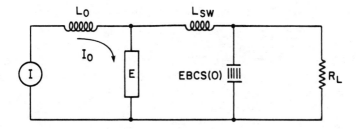

Fig. 9-6. Circuit diagram of an inductive storage system, illustrating the application of EBCS in the repetitive opening mode.

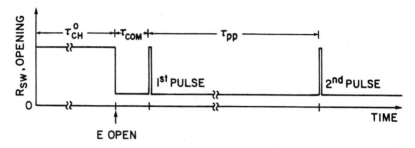

Fig. 9-7. Time behavior of the switch resistance of an EBCS(O) shown in Fig. 9-6. Reproduced from Ref. 6.

ductive system of Fig. 9-6. The EBCS(O) in Fig. 9-6 must conduct for the period between pulses, τ_{pp}, and open repetitively for a much shorter period, τ_p, represented by short spikes of high resistance. The output pulse shape, however, imposes stringent performance requirements on EBCS(O) in terms of pulse risetime and duration.

Although this scenario is conceptually the simplest, it may be too speculative to assume that a single switch can achieve an opening time of less than 10 ns and a conduction time of greater than 10 μs. Thus, this design has not been considered for detailed engineering. However, the option of using a purely inductive system is attractive and, therefore, suggests that future effort should be invested in development of switches with opening times of ~10 ns, (noting that change in the gas resistivity in times of <10 ns are theoretically possible[5], although an e-beam driver with the necessary waveform capability needs yet to be developed).

Electrical Characteristics for Opening Switch—Hybrid System. The energy associated with a single pulse for a single module of 20 kA, 200 kV, and 40 ns is very small: $E_1 = 160$ J. Considering a 10 pulse burst, ~2 kJ is required as a minimum to be stored by the inductor. Taking 30% efficiency of conversion from stored to pulse energy, ~6 kJ must be handled by a storage system for charging one pulse line module. This projects to hundreds of kilojoules for an accelerator consisting of many modules, for example, an ATA. In considering the engineering of a switching system, one must know the number of modules to be powered by one switch. A reasonable example that would uncover most of the problems of a design would be a switch for a 10 module pulse train generator limited to a burst of 5 pulses. Circuit, current, and voltage parameters are derived below for the EBCS(O) switch of Fig. 9-4.

Given the same final load pulse parameters as in Table 9-1, the single pulse line capacity is $C = 2E_1/V^2 = 8$ nF; i.e., in ten module operation 80 nF must be charged to 200 kV. If a 5 burst pulse train for an initial switch design is chosen, the total stored energy in the ten pulse lines is $E_2 = 8$ kJ. Assuming, further a 15 μs interpulse time (~ 70 kHz) and a 2 μs charge time, the energy stored in the inductor must be about $E_3 = 3 \times E_2 = 24$ kJ to account for the inefficiencies ($\sim 70\%$) associated with the circuit and fuses shown in Fig. 9-4. This amount of energy suggests that low voltage capacitors can be used conveniently as the source of current (shown in Fig. 9-4) to charge the inductor in small laboratory experiments. As previously stated, the time required to pulse-charge the capacitor C is taken to be $\tau_{CH} = 2$ μs. This is consistent with compact water-dielectric pulse line requirements[29]. The current needed to charge a capacitance to a given voltage, V, is $I_0 = dQ/dt = CV/\tau_{CH}$. For $C = 80$ nF, $V = 200$ kV and $\tau_{CH} = 2$ μs, $I_0 = 8$ kA. The storage inductor $L_0 = 2E_3/I_0^2 = 750$ μH. This choice of inductor presents no structural or electrical design problems.

To summarize, the EBCS(O) of Fig. 9-4 charging the ten pulse lines must accommodate the circuit performance characteristics shown in Table 9-2.

Switch Designs

Having established the electrical performance parameters of the switch by specifying the load characteristics and by selecting a specific circuit with its elements consistent with the requirements, it is now possible to incorporate the physical behavior of the switching gas (and the ionization-producing electron beam) to provide physical parameters of the switch design. In this section, the results of experimental and theoretical research and the switch performance criteria described previously are combined to obtain specific switch designs. First, the important switch characteristics and their relationship to the system are reviewed. A simple quantitative model for the switch physics along with scaling relations is then presented. A design pro-

Table 9-2. Summary of Assumed Circuit Characteristics. (Opening Switch, Hybrid System)

Type of load	Capacitive
Peak voltage, V_L	200 kV
Load current, I_0	8 kA
Conduction time, $\tau_C = (\tau_{pp} - \tau_{CH})$	13 μs
Non-conduction time, $\tau_{NC} = \tau_{CH}$	2 μs
Inductor, L_0	750 μH
Pulses per burst, n	5

cedure is outlined that self-consistently incorporates switch electrical characteristics with switch physics. Finally, the values of the switch parameters are obtained and the closing and opening switches are designed.

Switch Parameters. The switch parameters that determine opening and closing switch designs are switching gas breakdown, efficiency, gas resistivity during conduction, and closing or opening times. Specific role of each of them is now considered.

Breakdown. If the electric field across the switch exceeds the static breakdown field at the ambient switch pressure, the switch may go into an arc mode, which prevents the switch from opening as desired. This leads to the constraint that

$$V_L = s_1 \frac{E_B^0}{P_0} (P\ell) , \qquad (9\text{-}1)$$

where V_L is the maximum expected voltage across the switch (i.e., the load voltage), E_B^0 is the static breakdown field at atmospheric pressure P_0, s_1 < 1 is a dimensionless safety factor, and ℓ is the switch length. A linear relation of breakdown dependence on pressure is assumed, where $E_B = E_B^0 (P/P_0)$, with E_B being the static breakdown field at pressure P. This condition can be relaxed somewhat for transient pulses of less than 1 μs duration, as mentioned earlier.

Additionally, cummulative heating of the gas must be sufficiently constrained so that any local reduction in switch gas density does not significantly lower the self-sparking threshold. Energy is deposited in the switch from the following processes:

1. During the time the switch changes from a conducting to a nonconducting state, i.e., during the opening time τ_O, the current in the switch is finite while the switch resistance is large. Thus, the resistive heating during this time may be significant. This cummulative heating is estimated by

$$H_0 = k_1 I_{SW} V_L n \tau_0 \qquad (9\text{-}2)$$

where I_{SW} is the maximum switch current, V_L is the maximum switch voltage (load voltage), n is the number of pulses, τ_O is the opening time, and $k_1 \leq 0.5$ is a dimensionless constant which appropriately averages $(I_{SW} V_L)$ during τ_O. Because $\tau_0 \gg \tau_{SR}$, where the switch rise time, τ_{SR}, is the time for the switch to change from non-conducting to conducting, the energy loss during the switch closing phase is neglected.

2. During the total time the switch is conducting, $n\tau_0$, there will be some resistive heating. The total energy deposited in the switch as a result of this process is approximated by

$$H_C = I_{SW}^2 R_{SW} n \tau_C \qquad (9\text{-}3)$$

Here R_{SW} is the switch resistance during conduction.

3. When the beam is injected, a fraction of the beam's energy will be directly deposited in the switch gas as a result of inelastic collision processes. This energy deposition is estimated by

$$H_b = k_2 I_b V_b n \tau_C \qquad (9\text{-}4)$$

where I_b and V_b are the beam current and voltage respectively, and $k_2 \leq 1$ is the fraction of the beam energy deposited in the switch (determined from V_b, P, and ℓ). The conduction time τ_C is also the e-beam pulse duration.

To properly constrain cummulative heating in the switch, we must have

$$H_0 + H_C + H_b < W_B A \ell \qquad (9\text{-}5)$$

Here $W_B = W_B(E,P)$ is the deposited energy per unit volume at which the self-sparking threshold is reduced and A is the switch area. As shown in Fig. 9-8, W_B scales linearly with P (Ref. 11): $W_B = W_B^0 (P/P_0)$, where $W_B^0 \simeq 0.15$ J/cm^3 is the critical energy density at atmospheric pressure at which breakdown occurs[11, 30]. Equations (9-2)-(9-5) can thus be combined to give

$$[k_1 I_{SW} V_L \left(\frac{\tau_0}{\tau_C} \right) + I_{SW}^2 R_{SW} + k_2 I_b V_b] \, n\tau_C = \qquad (9\text{-}6)$$

$$s_2 \left(\frac{W_B^0}{P_0} \right) A \, (P\ell)$$

where $s_2 \leq 1$ is a dimensionless safety factor.

Efficiency. The switch energy gain, ξ, is defined as the ratio of the energy delivered to the load to the total energy used in making the switch

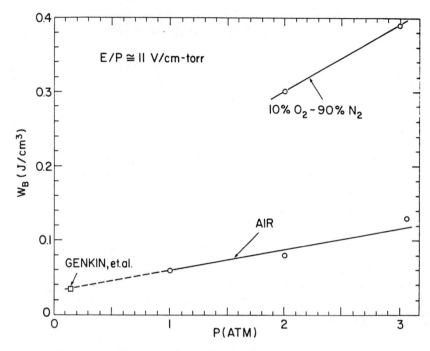

Fig. 9-8. Plot of deposited energy density, W_B, required for breakdown as function of ambient pressure, P, for air and a 10% O_2-90% N_2 mixture. Reproduced from Ref. 6. See Ref. 30 for datum point by Genkin et al.

conduct. Thus, applying the same arguments used in obtaining Eqs. (9-2)-(9-4), we have

$$\xi = \frac{I_L V_L \tau_L}{[k_1 I_{SW} V_L (\frac{\tau_0}{\tau_C}) + I_{SW}^2 R_{SW} + I_b V_b] \tau_C} \tag{9-7}$$

where I_L is the load current and τ_L is the load pulse width (e.g., $\tau_L = \tau_{CH}$ for the opening switch of the hybrid system in Fig. 9-4, while for closing switch $\tau_L = \tau_P$). In all cases, $I_{SW} = I_L$.

To attain a high efficiency,

$$\xi > 1 \tag{9-8}$$

Upon substitution of $I_L = I_{SW}$ and based on the definition of current gain, $\epsilon \equiv I_{SW}/I_b$, Eq. (9-7) becomes

$$\xi = \frac{(\tau_L/\tau_C)\,\epsilon}{[(\tau_0/\tau_C)k_1 + (I_{sw}R_{sw}/V_L)]\,\epsilon + 1} \qquad (9\text{-}9)$$

Measurements of ϵ as a function of percent O_2 in N_2 for 1, 2, and 3 atm at an applied $E/P = 10.5$ V/cm-torr with an e-beam current density of 5 A/cm² are illustrated in Fig. 9-9. Because of the nature of the e-beam pulse used in these measurements, the values of $\epsilon \le 15$ at low O_2 concentrations ($<20\%$) should be considered a lower bound.[11]

Equation (9-7) may be rearranged to yield

$$\xi^{-1}I_L V_L \tau_L = k_1 I_{sw} V_L \tau_0 + I_{sw} V_L \tau_c + I_b V_b \tau_C \qquad (9\text{-}10)$$

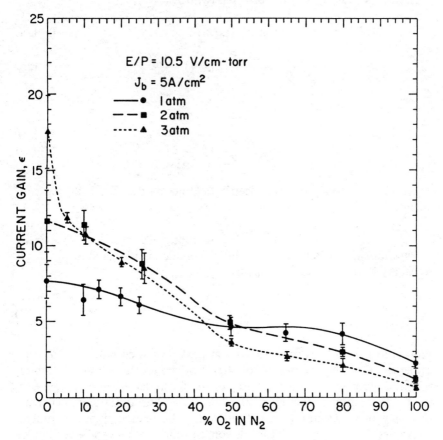

Fig. 9-9. Plot of current gain, ϵ, as a function of O_2 concentration in N_2 for $E/P = 10.5$ V/cm-torr and $J_b = 5$ A/cm² for pressures ranging from 1 to 3 atm. Reproduced from Ref. 6.

For the realization of an energy gain, i.e., for $\xi > 1$, each term on the right-hand side of Eq. 9-10 must be sufficiently less than $I_L V_L \tau_L$. Thus:

$$k_1 I_{SW} V_L \tau_0 = g_0 I_L V_L \tau_L \qquad (9\text{-}11a)$$

$$I_{SW}^2 R^{SW} \tau_C = g_C I_L V_L \tau_L \qquad (9\text{-}11b)$$

and

$$I_b V_b \tau_c = g_b I_L V_L \tau_L \qquad (9\text{-}11c)$$

where $g_0, g_C, g_b < 1$, such that $\xi^{-1} = g_0 + g_C + g_b < 1$.
Equations 11a–11c can be rewritten (with $I_{SW} = I_L$) as:

$$\tau_0 = (g_0/k_1)\,\tau_L \qquad (9\text{-}12a)$$

$$\frac{E_C}{P} = g_C s_1 \,(E_B^0/P_0)\,\frac{\tau_L}{\tau_C}, \qquad (9\text{-}12b)$$

and

$$\epsilon = g_b^{-1}\frac{V_b}{V_L}\left(\frac{\tau_C}{\tau_L}\right), \qquad (9\text{-}12c)$$

where $E_C = I_{SW} R_{SW}/\ell$ is the electric field across the switch during conduction.

Resistance.   Switch resistance during conduction is related to switch gas resistivity during conduction, ρ_0, by

$$R_{SW} = \rho_0 \frac{\ell}{A} \qquad (9\text{-}13)$$

Plotted in Fig. 9-10 is the resistivity at peak switch current as a function of percent O_2 in N_2. Values of 300–400 Ω-cm are typical for the concentration of oxygen $<20\%$ for injected electron beam density and applied electric fields specified in the figure. Because of the short duration (200 ns) and relatively long risetime (100 ns) of the electron beam used in the experiment, the values of ρ_0 in Fig. 9-10 for low concentrations of O_2 can be considered as an upper bound[11].

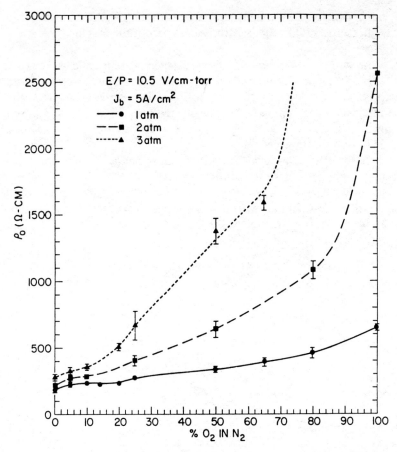

Fig. 9-10. Plot of resistivitiy at peak switch current, ρ_0, as function of O_2 concentration in N_2 at E/P = 10.5 V/cm-torr and J_b = 5 A/cm^2 for pressure range of 1-3 atm. Reproduced from Ref. 6.

Since $E_C = I_{SW}R_{SW}/\ell$ and Eq. 9-1 can be used to eliminate ℓ, Eq. 9-12b becomes

$$R_{SW} = g_C \left(\frac{V_L}{I_{SW}} \right) \frac{\tau_L}{\tau_C} \qquad (9\text{-}14)$$

for switch resistance during conduction. This equation and the condition $g_C < 1$ is essentially equivalent to the requirement that the characteristic

L/R time of the system be long compared with the switch conduction time, so that the current will not resistively decay from the system.

Closing and Opening Times. The characteristic time scale for the switch to change from a non-conducting to a conducting state is the switch *closing* time, τ_{SR}. This time is important in closing switch designs and typically in experiments has been limited by the beam risetime (~ 100 ns) and the circuit parameters. With the use of fast rising (<5 ns) beams, risetimes as short as 2 ns have been observed.[18]

The characteristic time scale for the switch to change from a conducting to a non-conducting state is the *opening* time, τ_0. On this time scale, the switch current decreases and switch resistance increases by orders of magnitude. Values of τ_0 obtained from single pulse experiments[9, 10, 11] are plotted in Fig. 9–11 as a function of the O_2 concentration in N_2 at 1, 2, and 3 atm with an applied $E/P = 10.5$ V/cm-torr for an e-beam current density of 5 A/cm². For this plot, τ_0 is estimated from the inductively generated

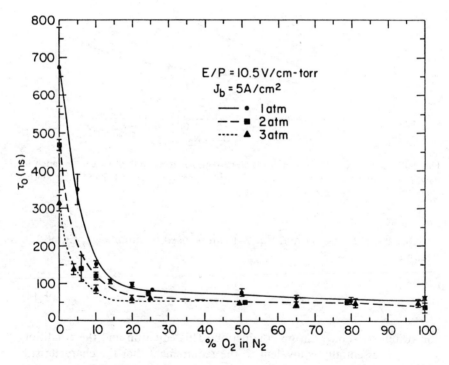

Fig. 9–11. Dependence of the EBCS opening time, τ_0, on O_2 concentration in N_2 at E/P and j_b values shown in the figure for three different values of pressure. Reproduced from Ref. 6.

voltage, V_1, i.e., $\tau_0 \approx L\Delta I/V_1$, where ΔI is the change in the system current and L is the system inductance. Typically, values of $\tau_0 \leq 300$ ns, limited by the beam decay time (~ 100 ns), are observed.

Switch Physics

The continuity equation for the switch plasma electron density, n_p, can be expressed simply as[11]

$$\frac{dn_p}{dt} = S_0 P J_b - \frac{n_p}{\tau_n} \tag{9-15}$$

where $J_b \equiv I_b/A$ is the e-beam current density and τ_n is the characteristic loss time for the switch plasma electron density. Here P is the pressure and S_0 is given by Eq. 9-17, below. Many τ_n periods are necessary for the switch to open. Specific number depends upon the dominant mechanism responsible for switch plasma electron loss. Thus we have

$$\tau_0 = k_3 \tau_n, \tag{9-16}$$

where $k_3 \approx 5$ for an attachment dominated switch or $k_3 \approx 10^2 - 10^3$ for a recombination dominated switch[5]. S_0 is an ionization parameter given by

$$S_0 = \frac{1}{e\,\epsilon_i P_0}\left(\frac{dE}{dX}\right)_0 \tag{9-17}$$

where e is the electronic charge, $\epsilon_i \simeq 35$ eV is the energy required for ionization per electron-ion pair, and $(dE/dX)_0 \simeq 3$ keV/cm is the energy lost per unit length for the beam electrons at 1 atm.

The plasma density is related to the switch plasma current density through

$$J_{SW} \equiv I_{SW}/A = n_p e v \tag{9-18}$$

where v is the electron drift velocity;

$$v = \mu E_C \tag{9-19}$$

Here μ is electron mobility[31] and E_C is the electric field across the switch. Thus resistivity during conduction is given by

$$\rho_0 = \frac{E_C}{J_{SW}} = (e\,n_p\mu)^{-1} \tag{9-20}$$

With the substitution of n_p from Eq. 9-20 into Eq. 9-15, and at equilibrium $dn_p/dt = 0$, Eq. 9-17 becomes

$$J_b \rho_0 \tau_n = f_0^{-1} \qquad (9-21)$$

where $f_0 \equiv eS_0 \mu P$ is essentially a constant for a given gas composition. For most gases, $f_0 \sim 10^5 - 10^6$ cm/V-s. Finally, with the use of the definitions of ϵ, J_{SW}, and J_b, Eq. 9-21 can be expressed as

$$\frac{\epsilon}{\tau_n} = E_c f_0 \qquad (9-22)$$

The relations derived in this section are used to relate switch physics to the switch circuit characteristics outlined above under "Switch Parameters."

Design Procedure

In this section, the switch circuit requirements are combined with the switch physics scaling laws to provide a self-consistent procedure for obtaining the switch gas composition and pressure, switch length, and switch area (radius) for a given switch gain.

First, the factor k_1 in Eq. 9-2 is obtained, which when multiplied by the power delivered to the load gives the average power dissipated by the switch during the opening phase. During the time the switch is undergoing a transition from conducting to a non-conducting state, the switch current is decreasing and the switch resistance is increasing. During τ_L, the switch current changes by I_{SW} ($= I_L$) and switch voltage changes by V_L; thus we define

$$k_1 = \frac{<I_{SW}> <V_{SW}>}{I_{SW} V_L} \qquad (9-23)$$

where $<I_{SW}>$ and $<V_{SW}>$ are the average values of the switch current and voltage during τ_0.

The time history of the voltage across the switch will depend upon the load voltage time behavior. For loads in which the load voltage rises to its peak value in a time $\simeq \tau_0$ (i.e, resistive or inductive loads, $<I_{SW}> \simeq \frac{1}{2} I_{SW}$ and $<V_{SW}> \simeq \frac{1}{2} V_L$ so that

$$k_1 \simeq \frac{1}{4} \qquad (9-24)$$

For a capacitive load (e.g., that seen by EBCS(O) in Fig. 9-4), the load voltage reaches its maximum in a time $\tau_L = \tau_{CH}$. In this case, we still have

$<I_{SW}> \simeq \frac{1}{2}I_{SW}$. However, $<V_{SW}> \simeq <I_{SW}> \tau_0/2C = V_L \tau_0/4\tau_L$. Thus, for a capacitive load,

$$k_1 \simeq \frac{1}{8}(\tau_0/\tau_L) \tag{9-25}$$

Once k_1 is known, the opening time, τ_0, can be computed from Eq. 9-12a for a choice of g_0. With τ_0 known, Eq. 9-16 can be used along with data on attachment and recombination rates to make a judgement on whether the switch should be attachment or recombination dominated, thus suggesting a specific gas composition. If τ_0 is not consistent with circuit requirements, then a different value of g_0 must be chosen.

The switch pressure is computed from Eq. 9-12b by substituting E_C from Eq. 9-22, ϵ from Eq. 9-12c, τ_n from Eq. 9-16 and τ_0 from Eq. 9-12a. Thus

$$P^{-1} = \frac{g_0 g_c g_b s_1 f_0}{k_1 k_3} \tau_L \frac{E_B^0}{P_0} \left(\frac{V_L}{V_b}\right) \left(\frac{V_L}{\tau_C}\right)^2 \tag{9-26}$$

The g factors must be chosen to be consistent with the inequality which follows Eq. 9-11c and can be optimized using the method of Lagrange undertermined multipliers, with the result that the three factors are about equal. For example, if $\xi = 3$ is chosen, $g_0 = 0.1$ and the switch efficiency $\eta = \xi/(\xi + 1) = 0.75$.

The switch length, for given P, is obtained from Eq. 9-1:

$$\ell = \frac{V_L}{s_1 P (E_B^0/P_0)} . \tag{9-27}$$

Combining V_b (in Eq. 26) with P and ℓ, gives k_2 in Eq. 9-4 to provide the value of energy deposition into the switching gas. The electron beam current can be obtained from Eq. 9-12c using the definition for current gain, ϵ:

$$\epsilon \equiv I_{SW}/I_b = g_b^{-1} \frac{V_b}{V_L} \left(\frac{\tau_C}{\tau_L}\right) \tag{9-28}$$

Substituting, further, the value of R_{SW} (from Eq. 9-14) and I_b from Eq. 9-28 into Eq. 9-6 provides the required switch area, A. This value of A insures that the switch will be sufficiently large so that the total deposited energy will not lower appreciably the breakdown threshold. The area is

$$A = \frac{I_{SW} V_L (n\tau_L) g_c}{s_2(W_B^0/P_0)P\ell} \left[1 + \frac{k_1}{g_c} \left(\frac{\tau_0}{\tau_L}\right) + k_2 \frac{g_b}{g_c}\right] \tag{9-29}$$

Design For Repetitive Closing Switch—Capacitive System. Taking the circuit parameters described in Table 9-1, the results of the last section ("Design Procedure") enable us to arrive at a switch design for the repetitive closing switch of Fig. 9-2. In this case, the parameters that are chosen are for repetitive switching operation which at this time cannot be achieved by any switches developed so far. For illustration, $\tau_{pp} = 15$ μs, safety factors $s_1 = 0.75$, $s_2 = 0.5$, and $V_b = 200$ kV. As previously stated (Table 9-1) $I_{SW} = I_L = 20$ kA, $n = 5$, $V_L = 200$ kV, and $\tau_p = 40$ ns. Choosing $\xi = 2$ gives $g_0 \simeq g_C \simeq g_b = 0.17$. With $\tau_C = \tau_p = \tau_L = 40$ ns, Eq. 9-28 requires $\epsilon \simeq 6$. Because the load is not capacitive, we substitute k_1 from Eq. 9-24 into Eq. 9-14a to compute τ_0, i.e., $\tau_0 \simeq 4g_0\tau_L \simeq 30$ ns. It is not disturbing that $\tau_0 \sim \tau_L$ because the pulse line does most of the load pulse shaping; therefore, the major requirements are that $\tau_0 < \tau_{pp}$ and that $\tau_{SR} \le \tau_R$. The second requirement can be easily met for an e-beam risetime $\le \tau_n$ (Ref. 18). We chose k_3 in Eq. 9-16 to be ~ 5 (attachment dominated) so that the plasma electron decay time $\tau_n \approx 6$ ns. This can be achieved with a N_2-O_2 gas mixture of 4:1 (Ref. 5).

The switch pressure is given by Eq. 9-26 with $E_B^0 \simeq 20$ kV/cm, $f_0 = 2 \times 10^5$ cm/V-s for N_2-O_2 and $k_1 = 0.25$ (Eq. 9-24), so that $P \simeq 1710$ torr $\simeq 2.3$ atm. The switch length, ℓ, is then, from Eq. 9-27, approximately 5.9 cm. These ℓ, P, and V_b values are consistent with $k_2 \simeq 0.3$. Equation 9-29 is then used to obtain the switch area, $A \simeq 2589$ cm^2, and radius, $r \simeq 29$ cm, giving a switch current density of $J_{SW} \simeq 8$ A/cm^2. The e-beam current density is thus $J_b \simeq 1.3$ A/cm^2. From Eq. 9-14 we compute $R_{SW} \simeq 1.7$ Ω and from Eq. 9-13 we obtain $\rho_0 = 745$ Ω-cm. Table 9-3 is a summary of the values of the switch parameters, which, along with the required switch inductance (≤ 100 nH), completely characterize the switch.

Table 9-3. Characteristics of Closing Switch for a Capacitive Pulser.

Switch gas composition	N_2:O_2 $\simeq$ 4:1
Switch pressure, P	2.3 atm
Switch length, ℓ	5.9 cm
Switch area (radius), $A(r)$	2589 cm^2 (29 cm)
Switch current density, J_{SW}	8 A/cm^2
Switch energy gain (current gain), $\xi(\epsilon)$	2(6)
Switch resistance during conduction	1.7 Ω
Switch voltage drop during conduction	34 kV
e-beam voltage, V_b	200 kV
e-beam current, I	3.5 kA
e-beam risetime $\le \tau_n$	≤ 2 ns
e-beam decay time $\le \tau_0$	≤ 30 ns

The e-beam generator is required to deliver ≥ 20 J/pulse with a very fast rise and decay time. The e-beam current density, however, is modest, ~ 1 A/cm^2.

Design For Repetitive Opening Switch—Hybrid System.   Taking the circuit parameters described in Table 9-2 and again following the prescription of "Design Procedures" the repetitive opening switch design for EBCS(O) of Fig. 9-4 is obtained. For this case $\tau_{pp} = 15$ μs, $\tau_{NC} = \tau_L = 2$ μs, $I_{SW} = I_L = I_0 = 8$ kA, $n = 5$, and $V_L = 200$ kV (Table 9-2). As before, $s_1 = 0.75$, $s_2 = 0.5$, and $V_b \simeq V_L = 200$ kV. Setting $\xi = 4$ gives, according to the condition for ξ^{-1}, $g_0 \simeq g_C \simeq g_b = 0.08$. With $\tau_C \simeq \tau_{pp} - \tau_L$, Eq. 9-28 requires $\epsilon \simeq 80$. Because of the capacitive load, k_1 from Eq. 9-25 is substituted into Eq. 9-12a to compute τ_0. Thus $\tau_0 = \sqrt{8g_0}\,\tau_L \simeq 1.6$ μs. Since $\tau_0 \sim \tau_L$, a substantial fraction of the final charging voltage for the capacitor will be attained during the opening time. This presents no serious problems, because the major requirement is that the capacitor be charged in a time $\leq \tau_{pp}$. Choosing k_3 in Eq. 9-16 to be ~ 5 (for the attachment dominated case) so that $\tau_n \simeq 300$ ns, which can be readily attained using N$_2$ with a small ($\sim 1\%$) admixture of O$_2$.

The switch pressure is given by Eq. 9-26 with $f_0 = 2 \times 10^5$ cm/V-s for N$_2$ and $k_1 = 0.1$ (Eq. 9-25), giving $P = 5226$ torr $\simeq 7$ atm. The switch length is then (from Eq. 9-27) 1.9 cm. Equation 9-29 is then used to obtain the switch area, $A = 3004$ cm^2, and the radius, $r = 31$ cm, giving a switch current density of $J_{SW} \simeq 2.6$ A/cm^2. The e-beam current density is thus $J_b \simeq 0.03$ A/cm^2. At this point, the switch resistance can be computed as in the preceeding section.

Table 9-4 summarizes the values of the switch parameters, which, along

Table 9-4. Characteristics of Opening Switch for a Hybrid Pulser.

Switch gas composition	N$_2$:O$_2$ $\sim$ 99:1
Switch pressure, P	7 atm
Switch length, ℓ	1.9 cm
Switch area (radius), $A(r)$	3004 cm^2 (31 cm)
Switch current density, J_{SW}	2.6 A/cm^2
Switch gain (current gain), $\xi(\epsilon)$	4 (80)
Switch resistance during conduction, R_{SW}	0.3 Ω
Switch voltage drop during conduction	2.4 kV
e-beam voltage, V_b	200 kV
e-beam current, I_b	100 A
e-beam rise and decay time	$\leq$ 100 ns

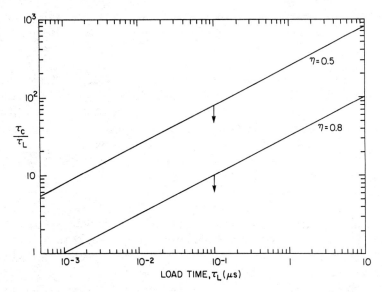

Fig. 9-12. Relationship of the conduction time, τ_c (normalized to the load time, τ_L) to the load time based on Eq. 9-26 for switch efficiency, η, of 0.5 and 0.8. Reproduced from Ref. 6.

with a required switch inductance (~ 100 nH), completely characterize the switch.

The e-beam generator requirements are quite modest with only 2.6 J/pulse needed. This reduces the complexity of the e-beam pulse modulation. The pressure requirement may be relaxed by as much as an order of magnitude via the choice of a gas with a higher mobility during conduction, e.g., a mixture of Ar with O_2 or CH_4 with an attaching gas (CO_2). Alternately, use of these gases will increase the switch efficiency already given in Fig. 9-12 if the switch pressure can remain high, as is evident from Eq. 9-26. More recent research in this area[23] is associated with the mobilities in gases that exhibit strongly (electric) field-dependent mobility such as that of 20% O_2-80% Ar and 99% CH_4-1% C_2F_6, which contrast with the normal behavior of 20% O_2-80% N_2 and 100% CH_4 respectively.

REMARKS RELATED TO EBCS

There are three principal objectives in the discussion above. The first is the review of the principles of operation of electron-beam controlled switches, emphasizing the ability of these switches to recover and open rapidly under high applied voltage. This particular point is made to emphasize the rather unique capability that makes the EBCS attractive, especially when compared with other switch candidates, either as a single pulse opening switch or as a switch for repetitively pulsed systems with high repetition rates.

The second objective is to develop the formalism of switch design. The formalism goes beyond the qualitative discussion of principles of opening and repetitively operated switches of Chapter 2. It emphasizes the overall energy transfer efficiency of the switch, the fundamental circuit requirements, and the scaling arising from switch physics. The analysis indicates that efficiencies of ~80% should be achievable for the examples chosen. These efficiencies are conservative in that the switch designs utilize the well known but non-optimum gases N_2 and O_2, as for example deriving the scaling shown in Fig. 9-12. For these gases, the product $\rho_0 \tau_0$ is constant[11]. Significant improvements in switch energy gain should result if N_2 is replaced by a gas such as CH_4 with high electron mobility, or if O_2 is replaced by a gas such as C_2F_6 in which the attachment rate increases rapidly with the applied field[23]. Such an increase in the attachment rate is especially strong at higher pressures, as can be seen from Ref. 23.

The third objective of the above detailed treatment is to illustrate the capabilities of the EBCS in a parameter regime of interest, both as an opening switch for inductive energy store and as a fast closing switch. In the examples chosen, the opening switch will conduct for ~15 μs and open in ~1 μs; the closing switch would close in ~5 ns and conduct for ~50 ns. The switch size in both cases is roughly one-half meter in diameter by a few cm in length.

Although single pulse operation has not been stressed here, it is important to note that the EBCS is capable of providing opening times substantially shorter than what can be achieved with wire fuses (discussed in Chapter 8), the only alternative available for fast opening applications. New developments in erosion switches, described next in this chapter, may provide a fast opening time, but the conduction times may be limited to very short durations. Therefore, the EBCS can be expected to make a significant contribution to the general development of inductive storage, if successful practical packaging can be developed. One such design has been discussed by Nolting et al.[33] This design and its integration into a pulse line are shown in Fig. 9-13. Nolting suggests that to obtain repetive switching at theoretically possible rates which may be as high as 2 MHz, it will be necessary to use thermionic emission cathodes as the source of the electron beam.

Thermionic electron tubes capable of producing current densities of 1A/ cm^2 are presently available commercially. With the use of a close-proximity grid, the beam source can be actuated with about a 100 V grid pulse. This gives the possibility of controlling a several hundred kilovolt switch with a standard solid state trigger generator. Figure 9-13a is a conceptual design

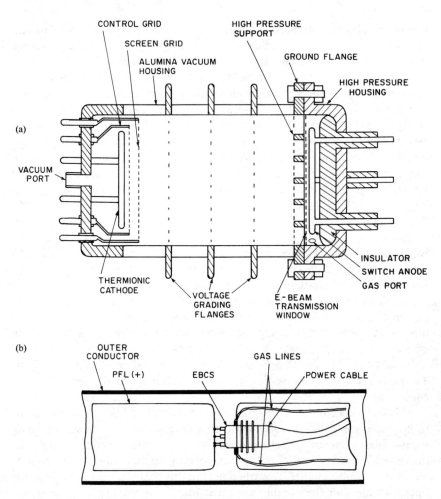

Fig. 9-13. Diagram of a compact EBCS design suggested by G. Nolting[33]. Also shown is the packaging of the switch into a high power pulse line for repetitive operation.

of how the thermionic housing would depend upon the operating requirements of the electron beam. A housing 15 cm long should be sufficient for switch voltages of 200 kV, making it a relatively compact tube. Such compact configuration of the EBCS can be easily incorporated into the coaxial geometry of conventional, high-power pulse lines. This is also illustrated in Fig. 9-13b, showing how the EBCS can be installed in a coaxial line and used to discharge the pulse forming line. The operation of the EBCS requires that the line be charged positive relative to ground, and since the electron source must be operated at negative potential, operating voltages can be maintained at reasonable levels.

Because EBCS has not yet been developed into an operating system, experience suggests that there may be other issues of concern to EBCS design and operation. For example, the impact of electrode sheath effects, current conduction by ions, or vacuum interface problems have not been discussed. Nonetheless, none of these additional issues should seriously compromise the design performance. This statement is supported by experiments performed at NRL and elsewhere[1, 8, 11, 34, 35, 36].

Finally, it is suggested by Bychkov et al.[37] that the electron beam used as an ionization source can be replaced by a strong ultraviolet source. They describe conditions under which gases with small amounts of electronegative molecules can be ionized to an electron density level of a few times 10^{12} cm^{-3} at about 1 atm. A very effective UV source would be a dielectric flashover generated by a low current (<100 A) 8–20 kV pulser. The flashover surface in Ref. 37 serves as a (plasma) cathode. The profiled anode is some 2.6 cm away. The main discharge is controlled by the gas in the interelectrode region. The switched currents and voltages are in the range of 100 A and 10 kV for 50 cm^2 discharge areas. Even though the main objective of the authors was to develop a simple method for maintaining stable volume discharge in lasing gases (CO_2:N_2:He mixture), they note that the plasma cathode can switch on and off rapidly at high power, at frequency of about 50 kHz.

EROSION SWITCHES

Erosion switches (PEOS) are a new type of switches operating on a nanosecond timescale. They were initially developed for pulse steepening or prepulse suppression[2, 38, 39, 40]. The latter function is employed to eliminate a precursor pulse appearing at the output of high power pulse lines as a result of capacitive coupling across the output switch, for example isolating the pulse line from the load. The typical conduction time of the erosion switch (PEOS) is limited to less than 100 ns and the opening time is about 10 ns.

The operation of the erosion switch begins with the plasma from an appropriate source, such as that developed by C. W. Mendel et al.[2], filling the switch region and connecting, in effect, the switch electrodes with a conducting medium. Figure 9–14 shows such a source in an experimental apparatus driven by a GAMBLE I pulse line. An equivalent circuit, containing an optional inductor is also shown. Application of voltage from a pulser across the (coaxial) electrodes (and across an inductor if the switch serves as a current interruptor for generation of an inductive voltage) leads to a current flow through the switch (and through the inductor if one is

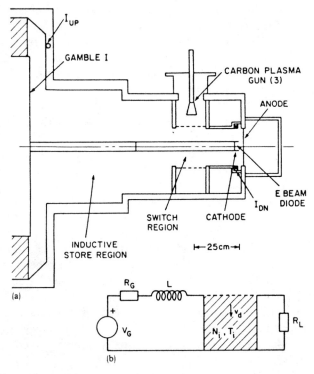

Fig. 9-14. Schematic of the plasma opening switch in experiments performed at the Naval Research Laboratory. Also shown is the circuit diagram of the experiment where R_G is the 2 ohm pulse line generator impedance and V_G is its open circuit voltage. L_S and R_L are the inductor and load impedance respectively. Reproduced from Ref. 40.

used). As Meger et al. indicate, a cathode surface plasma is formed and becomes a space-charge limited electron emitter[3, 41]. A gap near the cathode plasma is formed and a bipolar Child-Langmuir diode is established. Ions enter such a gap, formed by the cathode plasma and (carbon) plasma due to their drift and thermal velocity, supplying less than 1% of the charge flux necessary to maintain the bipolar flow. The ion flux necessary to maintain the current flow is

$$j_i/j_e = (M_e/2M_i)^{1/2} (\gamma + 1)^{1/2} \qquad (9\text{-}30)$$

(where m_e, M_i and $\gamma m_e c^2$ are the electron and ion rest masses and the maximum electron energy respectively)[42, 43]. The total bipolar current density is

$$j = j_e + j_i = (1 + j_i/j_e) \times 4.3 \times 10^{-6} V^{3/2} d^{-2} \qquad (9\text{-}31)$$

Here V is the gap voltage (in volts) and d is the gap thickness (in cm). The ions enter the gap due to thermal and drift velocity, and in the case of commonly used C^{+2} plasma, a charge flux of 20 A/cm² is sufficient to drive a bipolar current of 2 kA/cm². As long as the current density remains below this level, the gap impedance remains low.

Following further the model described in Ref. 40, a new phase begins when the ion flux into the gap is no longer maintaining the bipolar flow, as would happen when the current source continues to increase the current. An additional flux is then provided by eroding the plasma and opening the gap. The rate of erosion can reach 10 cm/μs. As the gap, d, increases, the Child-Langmuir switch impedance increases with d^2 giving a rapidly increasing switch resistance. As the current in the switch region increases, the switch electrons are bent downstream by their self-magnetic field and travel along the plasma surface for the length, L, of the switch plasma.

This begins the next phase of the opening process. The ion flux and consequently the erosion rate increases by a factor of L/d in the same way the ion current in a pinched-beam ion diode[42, 43] is enhanced. Opening velocities of >100 cm/μs are possible with this enhancement, potentially leading to very fast increases in the opening of the switch. As the switch resistance increases, the load begins to conduct a larger fraction of the current. With the series inductor, the system acts like a constant current source, increasing the voltage across the switch. In such experiments, the output (switch) voltage can be a few times that of the pulse line charging the inductor.

The final phase begins when the diode current exceeds the critical current for magnetic insulation in the switch gap. The electrons no longer interact with the plasma and travel down the final section of transmission line to the load. When the load current exceeds the critical current, $I_c = 8.5 \times 10^3 \beta \gamma \, R/d$ (in Amperes), where β is the normalized electron velocity at full voltage, the electrons become magnetically insulated[40].

During the development of erosion switching techniques and in determining the limitations of erosion switching in general, a variety of switching modes were tested. The authors in Refs. 40 and 41 describe experiments in which prepulse elimination and pulse steepening were demonstrated and increased output voltage was achieved. The experimental hardware consisted of a 40 cm long, 100 Ω coaxial vacuum inductor, a switch section, and an e-beam diode section. The ~ 140 nH, high impedance vaccum inductor section and the ~ 35 nH insulator provided the $L_S \simeq 175$ nH inductive store. The interruption of the current flow in the inductor was performed in the switch section containing three plasma guns of a type described in Ref. 2. The guns were located 12 cm off axis and injected a carbon plasma (primarily C^{+2}) through a 10 cm diameter metal screen, striking the inner 5 cm diameter cathode surface over a ~ 60 cm² area. The plasma ion density was measured to be $\sim 1.5 \times 10^{13}$ cm^{-3} at peak flux. It moved with a drift velocity of 7.5 cm/μs. Neutrals from the guns had a velocity of approximately 1 cm/μs and arrived after the plasma. Downstream of the switch region was a 10 cm long, 42 Ω transmission line segment with inductance (L_L) = 14 nH, followed by the e-beam diode load. The diode consisted of a 6 mm thick rounded edge, hollow, cylindrical aluminum cathode, typical for high power electron beam generators. The cathode was located 1 cm from a 1.6 mm thick carbon coated aluminum anode.

Shown in Fig. 9–15 is experimental data from two shots[41], comparing behaviors with and without the erosion switch. The shots had nearly identical peak pulser voltage of -960 kV across the inductive store section. The pulse was applied at the inductor 0.5 μs after the peak of the injected plasma flux at the cathode surface. The currents in the inductor, I_{UP}, and in the load section (an electron beam diode), I_{DN}, are shown in Fig. 9–15. Without the switch, the pulser saw 190 nH inductance ($L_S + L_L$) in series with the ~ 12 Ω e-beam load. This limited the peak current in the inductor to ~ 80 kA. I_{DN} follows I_{UP} with some current loss due to cathode stalk emission which is not measured by the current probe. With the switch, the pulse generator saw only the inductive store with $L_S = 175$ nH and the closed switch for the first ~ 50 ns. During this time, the switch diverts up to 220 kA from the load. When the switch begins to open, the downstream current rises to ~ 130 kA in ~ 6 ns with a maximum $dI_{DN}/dt \simeq 2.2 \times 10^{13}$ A/s. A loss of 90 kA occurs in the switch and downstream of the switch.

Voltages across the switch $V_{SW} = V_D - L_S dI_{UP}/dt$ computed from the measured insulator voltage V_D and dI_{UP}/dt are also shown in Fig. 15. The voltage across the load impedance is approximately the same as the switch voltage because $L_L \ll L_S$. The effect of the switch on the voltage risetime and amplitude is well demonstrated in these experiments. Without the switch, voltage peaks at about 790 kV and has an 85 ns FWHM (full width at half-maximum). With the switch, the voltage is near zero during the conduction phase, then rises to 1.4 MV in some 20 ns. The FWHM of the pulse is ~ 25 ns. The FWHM of the voltage pulse is reduced by a factor of three and the peak voltage is a factor of two higher than without the switch. The use of the switch also increases output power. The peak power into the switch and load combination is 0.065 TW without the switch and 0.28 TW with the switch. The four-fold inrease in peak power results from the higher voltage and current in the switch. For comparison, the power delivered to a matched (2 Ω) low inductance (30 nH) diode is 0.14 TW for a similar pulser output.

Practical operation of the switch has also been studied to determine the degree of plasma flow control necessary for a given operation[40, 41]. Switching performance data were taken with different plasma densities in the

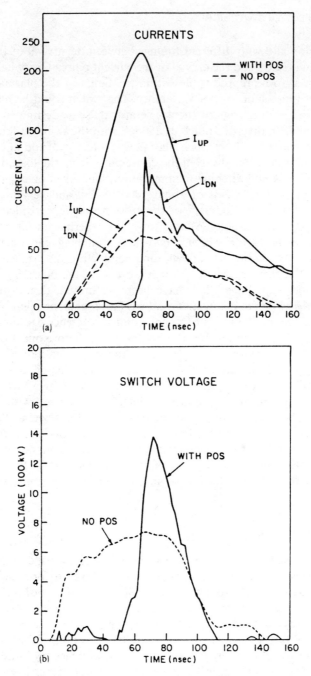

Fig. 9–15. Effect of the erosion switch opening on current and voltage, compared to the generator output without use of the switch. Reproduced from Ref. 40.

switch region and with different timings between the arrival of the plasma in the switch region and the arrival of the current pulse. The switch opening was observed to be reproducible within the limits of the plasma gun and accelerator pulse shot-to-shot variability. The switch could be made to divert only the leading edge of the accelerator pulse as desired for prepulse suppression or to divert as much as 250 kA with the switch still opening in ~ 10 ns depending on how much plasma was injected. If too much plasma was injected, the switch diverted the entire pulse from the load; i.e., under those conditions no switching (commutation) had taken place.

Very high hold-off fields of about 1000 kV/cm and very fast opening times suggest that development of the erosion switch for wider operating ranges would be desirable. At this time, no experiments have been performed to demonstrate whether two-stage switching can be effectively achieved by combining a relatively long conducting first stage, such as a fuse, with the erosion switch. In order to extend the operating ranges of PEOS, initial experiments[51, 52] have been performed, demonstrating the ability of the switch to conduct current for more than one microsecond and to open subsequently in tens of nanoseconds. Reference 51 describes such a switch capable of conducting 120 kA for 400 ns before opening in 40 ns. When an electron beam diode was used as the load of an inductive storage, the inductive voltage developed by the switch across the diode reached 170 kV. This voltage exceeded the initial capacitor voltage by a factor of four. Two to three times longer conduction time, in comparison with the 400 ns time, was achieved in the experiments reported by Bugaev et al.[52], in which a current of 400 kA was interrupted in 100 ns and transferred from PEOS to the diode load. In these experiments, the switch plasma density was 10^{14} cm^{-3}, resulting in an effective rate of resistance increase of 10^9 Ω/s, allowing as much as 70% of the current in the storage to be transferred to the load.

THE PLASMADYNAMIC SWITCH

A class of pulse generators such as capacitor banks, homopolar generators, and explosive flux compressing devices is employed to provide relatively cheap energy in large output pulses on a time scale of tens of microseconds to seconds. For many physics experiments and for development of high energy density high power pulser technology, these pulses are too long. Power amplification using efficient inductive storage techniques is employed. This is achieved by the use of opening switches for pulse compression as discussed in the preceeding sections and in Chapter 2. However, to satisfy the need for providing pulse compression from tens of microseconds to about one microsecond or less, various experiments have been performed to obtain data and develop techniques which would commutate currents at very high (multi-megampere) level. Demeter et al.[44] have performed tests on a staged implosion of initially solid conducting foil which is heated sufficiently (by the current to be commutated to the load) to become a plasma. S. Seiler et al.[4] carried out a similar test employing a coaxial gun using a metal foil that converts into plasma to accumulate and concentrate the electromagnetic energy. D. R. Kania et al.[45] worked, on the other hand, with more massive (relative to current and pulse time) metalic slab (also accelerated by the magnetic field of the current to be commutated) which is heated by Joule dissipation only to a level where the material loses its mechanical strength. The switching slab is called a "magnetic gate" in Ref. 45. Although, strictly speaking it is not a plasmadynamic switch, its circuit function is similar as that of Ref. 4 and 44 and provides a conceptually simple portrayal of its operation.

"Magnetic gate" switching is illustrated in Fig. 9–16. Initially, current is made to flow from a source (capacitor bank) through a metal foil (shown in the figure by fine cross-hatching) positioned so that the magnetic field in the circuit can accelerate the foil along the heavy electrode channel (or along coaxial electrodes). A load chamber is positioned at some distance. It is represented schematically in the figure as a narrow slot terminated by a short-circuit "load," which may be another thin foil that (in coaxial geometry) would be imploded onto the axis of symmetry. As the magnetic gate is accelerated past the slot, the current is "switched" to flow into the slot. Figure 9–16, reproduced from Ref. 45 to illustrate typical parameters, shows the current through a 2 mm thick Al gate switch flowing in the slots behind the gate and current transfer to two slots (one is shown in the figure) at a distance of 0.9 and 2.6 cm from the original gate position. Current risetime in the slot (i.e., at the positions of first and second slots) is 1.5 and 1.1 μs, respectively, with deduced gate velocities of 0.6 and 1.6 cm/μs.

Because of the large current and current density flowing through the moving contacts of the gate, the entrance to the slot can easily be short-circuited by plasma. This problem has been addressed by P. J. Turchi[46], who considers details of the plasma distribution in the vicinity of the slot, and the resulting current flow. The magnetic pressure would be sufficient to sweep such a plasma into the slot shown in Fig. 9–16. If the surface of the wall at the end of the slot in Fig. 9–16 obstructs the expanding magnetoplasma in the slot, recompression of the plasma will occur, converting the kinetic energy of the flow back into magnetic energy. Because the magnetic energy density exceeds substantially the thermal energy density of the driving plasma and its conductivity is high, stagnation of the current-carrying plasma at the load forms a fast-rising magnetic pressure pulse act-

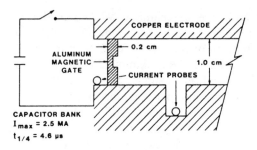

PLANAR GEOMETRY - 5 cm WIDE

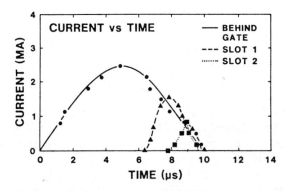

Fig. 9-16. Schematic drawing of the magnetic gate switch and typical current waveforms. Reproduced from Ref. 45. Current waveforms are obtained from magnetic probes indicated as small circles.

ing on the load. In the discussion of Ref. 46, this wall is replaced by a thin foil. Since cylindrical geometry is considered, the foil is imploded by the large magnetic pressure. In the largest tests to date, the SHIVA capacitor bank at the Air Force Weapons Laboratory[47] was used at the 3 MJ level (charged to 68 kV) to provide a peak current of 9.2 MA, rising in 3.5 μs. P. J. Turchi[48] reported for these tests a transfer of 8 MA to the implosion load (a foil of 10 cm diameter and 2 cm in length with a mass density of 164 μg/cm^2). The current risetime of 200 ns was achieved leading to a peak implosion speed of 20 cm/μs. The change in the load inductance, L, resulting from the compression suggests that the load voltage, $I(dL/dt)$, was about 0.5 MV.

Because of the lack of clear delineation of the current distribution in the driven plasma and in plasmas formed due to high current density at the electrodes, it is difficult to predict whether such plasmadynamic switches can be effectively used with loads other then some form of an implosion load. Discussion of the switching concept in Refs. 44, 45, 46 and in other publications by these authors suggest that the operation is not unlike that of the well researched plasma pinch, the so-called dense plasma focus[49].

To provide scaling relations that describe the behavior of the plasma-dynamic switch as an element of a circuit, W. H. Lupton[50] has derived the expressions for the switch current, I, the rate of change of inductance, dL/dt, associated with the motion of the plasma, the optimum switching time, t, and the maximum voltage V_{max} that appears across the current generator. To obtain such scaling, the switch plasma is highly idealized (by assuming infinite conductivity and thin-slab geometry) and the switching initially connecting the stationary plasma into the current generator is neglected. The generator is an inductor with inductance L_0. The plasmadynamic switch is described electrically as a conducting sheet connected to form a short circuit across the annular region between a pair of coaxial conductors as illustrated in Fig. 9-17. This annular sheet of plasma is accelerated by the magnetic force resulting from a current through the plasma and the coaxial conductors. Assuming no field penetration to the other side of the moving plasma, the equation of motion is

$$m(du/dt) = L_\ell i^2/2 \qquad (9\text{-}32)$$

where m = mass of moving conductor, u = velocity of moving conductor and

$$L_\ell = (\mu_0/2\pi)\, ln(r_2/r_1) \qquad (9\text{-}33)$$

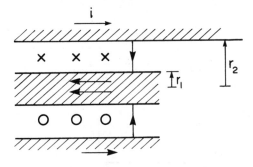

Fig. 9-17. Schematic of the coaxial plasmadynamic switch geometry.

is the inductance per unit length of the coaxial conductors, with radii r_1, and r_2 indicated in Fig. 9–17. Since the rate of change of inductance of the coaxial geometry is $dL/dt = L_\ell u$,

$$d^2L/dt^2 = (L_\ell^2/2m)i^2 \tag{9-34}$$

Current for the plasma flow switch is obtained from an energy storage inductor as shown in Fig. 9–18. Continuing with W. H. Lupton's[50] derivation, the inductor is considered to be initially energized with a current i_0. If the circuit resistance is neglected, then the total flux in the storage inductance, L_0, and switch inductrance, L, remains constant. In this case, the current in the plasma flow during the run-up (conducting) phase is

$$i = \frac{\phi_0}{L_0 + L} , \tag{9-35}$$

where the circuit flux, $\phi_0 = L_0 i_0$, is a constant. With this current, the equation of motion during run-up of the plasma flow is

$$\frac{d^2L}{dt^2} = \frac{L_\ell^2}{2m} \left(\frac{\phi_0^2}{(L_0 + L)^2} \right) \tag{9-36}$$

Upon multiplying both sides of the above equation by dL/dt and taking the initial condition to be $dL/dt = 0$ and $L = 0$ (the latter assumption is made for alegrabic simplicity; if the plasma flow begins with finite inductance, that value can be added to L_0) in order to evaluate the integration constant, the rate of change of inductance becomes

$$\frac{dL}{dt} = \frac{L_\ell}{\sqrt{m}} \frac{\phi_0}{\sqrt{L_0}} \sqrt{\frac{L}{(L_0 + L)}} \tag{9-37}$$

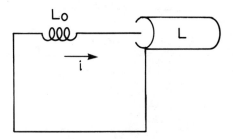

Fig. 9–18. Inductive energy storage circuit used in derivation of scaling relations (Eqs. 9–39, 9–40 and 9–45) for the plasmadynamic switch.

P. J. Turchi[48] has noted that, for good plasma flow switching, high velocity and high magnetic pressure is needed. This requirement can be translated into a figure of merit for the switch, i^2u. This figure of merit is proportional to[50]

$$i^2(dL/dt) = \frac{L_\ell}{\sqrt{m}} \frac{\phi_0^3}{\sqrt{L_0}} \left(\frac{L^{1/2}}{(L_0 + L)^{5/2}} \right) \qquad (9\text{-}38)$$

As L is varied, this expression has a maximum when $L = L_0/4$. Optimum switching is obtained by fixing the length of the plasma flow to achieve this value of inductance change. When the plasma flow reaches the end, the current will have fallen to $0.8i_0$ and

$$\frac{dL}{dt} = \frac{L_\ell}{\sqrt{m}} \frac{\phi_0}{\sqrt{L_0}} \frac{1}{\sqrt{5}} \qquad (9\text{-}39)$$

The above expression integrates to

$$t = \left(\frac{m^{1/2}}{L_\ell} \right) \left(\frac{L_0^{3/2}}{\phi_0} \right) \left[\frac{\sqrt{5}}{4} + \sinh^{-1}\left(\tfrac{1}{2}\right) \right] \qquad (9\text{-}40)$$

for the optimum switch performance (when $L = L_0/4$ is reached).

For a system with given L_0 and ϕ_0, a plasma dynamic switch can be made to achieve the optimum figure of merit after any conduction time, t, by selecting the geometric and mass parameter $L_\ell/m^{1/2}$ so that

$$\frac{L_\ell}{m^{1/2}} = \frac{L_0^{3/2}}{\phi_0 t} \left[\frac{\sqrt{5}}{4} + \sinh^{-1}\left(\tfrac{1}{2}\right) \right] = 1.04\, L_0^{3/2}/(\phi_0 t) \quad (9\text{-}41)$$

If the parameter is selected to have this value, then at the end of the conduction phase

$$\frac{dL}{dt} = \left[1/4 + \frac{\sinh^{-1}(1/2)}{\sqrt{5}} \right] \frac{L_0}{t} = 0.47 \left(\frac{L_0}{t} \right) \qquad (9\text{-}42)$$

Finally, the voltage across L_0 in this constant flux model is obtained from the condition

$$\frac{d\phi}{dt} = (L_0 + L)\frac{di}{dt} + i\frac{dL}{dt} = 0, \qquad (9\text{-}43)$$

so that

$$L_0 \, di/dt = \left(-\frac{L_0}{L_0 + L} \right) i \, (dL/dt) = -\frac{L_0}{\phi_0} i^2 \, (dL/dt) \qquad (9\text{-}44)$$

This is proportional to the figure of merit introduced earlier. Since the optimum switch maximizes this figure of merit at the end of its conducting run, the voltage for such a switch will also continually increase until the end of the run-up. At this time, $L = L_0/4$, $i = 0.8i_0$ and $dL/dt = 0.47$ L_0/t so the maximum voltage across the inductor is

$$V_{\max} = 0.3 \, \phi_0/t_0 \qquad (9\text{-}45)$$

The scaling, derived by W. H. Lupton[50], can be applied to a large storage system. To illustrate the procedure as well as to obtain typical parameters, a system with stored energy, $E_0 = 10$ MJ is considered. Further assumption that the inductive storage is "charged" with current $i_0 = 10$ MA dictates L_0, i.e., $L_0 = 2E_0/i_0^2 = 2 \times 10^{-7}$ H. The resulting flux is $\phi_0 = 2$ V-s and from Eq. 9-45 $V_{\max} = 0.6/t$ volts. To determine the run-up time t, Eq. 9-42 can be used: $dL/dt = 0.94 \times 10^{-7}/t$, with the above value of L_0. Assuming coaxial geometry, $L_\ell = dL/udt$, so that $L_\ell = 0.94 \times 10^{-7}/ut$. If the inner to outer conductor diameters are in a ratio of e, then from Eq. 9-33, $L_\ell = 2 \times 10^{-7}$ H/m. The final velocity u is determined by having the plasma flow cross a gap typically 2 cm in length. If this crossing (opening) time is to be 100 ns, the value of u is 2×10^5 m/s. Thus, $t = 2.7 \times 10^{-6}$ s (i.e., a ratio of conduction to opening time of 27). Finally, the maximum opening voltage is $V_{\max} = 220$ kV.

The description of the plasmadynamic switch in this chapter provides an introduction to very high energy switching for large inductive storage systems. The need of large experimental facilities for the development of this type of switch and the high cost of experiments has allowed for the provision of only very basic data so far. However, it can be seen already that the plasmadynamic approach could provide an effective way to accumulate and concentrate electromagnetic energy. For the successful operation of these switches, it may be particularly important to control tightly such design parameters as the distribution of current and mass density using multilayer foils[4]. This is a situation similar to that encountered with the electron beam controlled switch and the erosion switch; plasma switches for successful operation are much less tolerant to deviations in the choice of parameters and in their design compared to the closing switches and opening switches that depend on the cooling of the arc.

REFERENCES

1. B. M. Kovalchuk and G. A. Mesyats, "Rapid Cut-Off of a High Current in an Electron-Beam-Excited Discharge," *Sov. Tech. Phys. Lett* **2**, 252 (1976).

2. C. W. Mendel, Jr., and S. A. Goldstein, "A Fast-Opening Switch for Use in REB Diode Experiments," *J. Appl. Phys.* **48**, 1004 (1977).

3. R. A. Meger, J. R. Boller, R. J. Commissio, G. Cooperstein, Shyke A. Goldstein, R. Kulsrund, J. M. Neri, F. Oliphant, P. F. Ottinger, T. J. Renk, J. D. Shipman, S. J. Stephanakis, B. V. Weber, and F. C. Young, *Proceedings of the 5th International Conference on High Power Particle Beams,* San Francisco, CA, IEEE Cat. No. CONF-830911, p. 330 (1983).

4. S. Seiler, J. F. Davis, P. J. Turchi, G. Bird, C. Boyer, D. Conte, R. Crawdord, G. Fisher, A. Latter, W. Tsai, T. Wilcox, "High Current Coaxial Plasma Gun Discharges Through Structured Foils," *Digest of Technical Papers of the IEEE Fourth Pulsed Power Conference,* Albuquerque, NM, IEEE Cat. No. 83CH1908-3, p. 346 (1983).

5. R. F. Fernsler, D. Conte, and I. M. Vitkovitsky, *IEEE Trans. Plasma Sci.* **PS-8**, 176 (1980).

6. R. J. Commisso, R. F. Fernsler, V. E. Scherrer, and I. M. Vitkovitsky, "Design of Electron Beam Controlled Switches," Naval Research Laboratory Memorandum Report 4975 (1982). An abridged version of this work by the same authors appears in *Rev. Sci. Inst.* **55**, 1834–1840 (1984).

7. Use of PEOS (switch) for pulse steepening is described by R. A. Meger and F. C. Young, "Pinched-Beam Ion-Diode Scaling on the AURORA Pulser," *J. Appl. Phys,* **53**, 8543 (1982). Use of PEOS as a current interrupter for generating high inductive voltage is described by J. M. Neri, J. R. Boller, R. J. Commisso, R. A. Meger, P. F. Ottinger, T. J. Renk, B. V. Weber, and F. C. Young, "Improvement of PEOS Results on GAMBLE II," *IEEE 1985 International Conference on Plasma Science,* Pittsburgh, PA, IEEE Pub. 85-CH2199-8, p. 81 (1985).

8. L. E. Kline, "Performance Predictions for E-Beam Controlled ON/OFF Switches" *Proceedings of Workshop on Repetitive Opening Switches,* Tamarron, CO, Texas Tech Univ. Rept. p. 121 (1981).

9. R. J. Commisso, R. F. Fernsler, V. E. Scherrer, and I. M. Vitkovitsky, "Electron-Beam Controlled Discharges," *IEEE Trans. Plasma Sci.* **PS-10**, 241–245 (1982).

10. V. E. Scherrer, R. J. Commisso, R. F. Fernsler, L. Miles, and I. M. Vitkovitsky, "The Control of Breakdown and Recovery in Gases by Pulsed Electron Beams," *Gaseous Dielectrics III (Proceedings of the Third Symposium on Gaseous Dielectrics),* L. G. Christophorou, ed., Pergamon Press, New York, pp. 34–39 (1982).

11. V. E. Scherrer, R. J. Commisso, R. F. Fernsler, and I. M. Vitkovitsky, "Multiple Pulse Electron-Beam Controlled Switch," *IEEE Conference Record of the Fifteenth Power Modulator Symposium,* IEEE Cat. No. 82CH1785-5, pp. 146–152 (1982).

12. W. A. Barletta, Lawrence Livermore Laboratory Report UCRL-87288 (1981).

13. T. R. Burkes, J. P. Craig, M. O. Hagler, M. Kristiansen, and W. M. Portnoy, *IEEE Trans. Electr. Devices* **ED-26**, 1401 (1979). PEOS, not included in this reference, should also be considered for repetitive operation according to R. A. Meger, Naval Research Laboratory, in a private communication (1985).

14. E. J. Lauer and D. L. Birx, "Low Pressure Spark Gap," *Third IEEE International Pulsed Power Conference,* Albuquerque, NM, IEEE Cat. No. 81CH1662-6, pp. 380–383, (1981).

15. I. D. Smith, G. Lauer and M. Levine, "Tests of a Dielectric-Vacuum Surface Flashover Switch," *IEEE Conference Record of the Fifteenth Pulse Power Modulator Symposium,* IEEE Cat. No. 82CH1785-5, pp. 160–163 (1982).

16. D. L. Birx, E. G. Cook, L. L. Reginato, J. A. Schmidt, and M. W. Smith, "The Application of the Magnetic Pulse Compression to the Grid System of ETA/ATA Acceler-

ator," *Conference Record of 1982 Fifteenth Pulse Power Modulator Symposium,* IEEE Cat. No. 82CH1785-5, pp. 10–14 (1982).

17. D. L. Birx, E. J. Laurer, L. L. Reginato, D. Rogers, Jr., M. W. Smith, and T. Zimmerman, "Experiments in Magnetic Switching," *Third IEEE International Pulsed Power Conference,* Albuquerque, NM, IEEE Cat. No. 81CH1662-6, p. 262 (1981).

18. K. McDonald, M. Newton, E. E. Kunhardt, M. Kristiansen, and A. H. Guenther, "An Electron-Beam Triggered Spark Gap," *IEEE Trans. Plasma Sci.* **PS-8,** 181–185 (1980).

19. J. E. Eninger, "Broad Area Electron Beam Technology for Pulsed High Power Lasers," *Third IEEE International Pulsed Power Conference,* Albuquerque, NM, IEEE Cat. No. 81CH1662-6, pp. 499–503 (1981).

20. A. W. Friedman and J. E. Eninger, "Repetitively Pulsed Dispenser Cathodes," *Third IEEE International Pulsed Power Conference,* Albuquerque, NM, IEEE Cat. No. 81CH1662-6, pp. 519–522 (1981).

21. C. A. Spindt, I. Brodie, L. Humphrey, and E. R. Westerberg, "Physical Properties of Thin-Film Field Emission Cathodes with Molybdenum Cones," *J. Appl. Phys,* **47,** 5248 (1976).

22. B. Fell, R. J. Commisso, V. E. Scherrer, and I. M. Vitkovitsky, "Repetitive Operation of an Inductively-Driven Electron Beam Diode," *J. Appl. Physics* **53**(4), pp. 2818–2824 (1982).

23. V. E. Scherrer, R. J. Commisso, R. F. Fernsler, and I. M. Vitkovitsky, "Study of Gas Mixtures for E-Beam Controlled Switches," *Gaseous Dielectricts IV, Proceedings of the Fourth Symposium on Gaseous Dielectrics,* L. G. Christophorou and M. O. Pace, eds., Pergamon Press, New York, pp. 238–245 (1984).

24. L. A. Miles, E. E. Nolting, I. M. Vitkovitsky, and D. Conte, "Design of Large Area E-Beam Controlled Switch," *IEEE Conference Record, 1980 Fourteenth Pulse Power Modulator Symposium,* IEEE Cat. No. 80CH1573-5ED, pp. 68–72 (1980).

25. N. W. Harris, F. O'Neill, and W. T. Whitney, "Compact High Pressure Electron-Beam-Controlled Laser System," *Rev. Sci. Inst.* **48,** 1042–1049 (1977).

26. D. Conte, R. D. Ford, W. H. Lupton, and I. M. Vitkovitsky, "Two-Stage Opening Switch Techniques for Generation of High Inductive Voltages," *Proceedings of the Seventh Symposium on Engineering Problems in Fusion Research,* Knoxville, TN, IEEE Pub. 77CH1267-4-NPS, pp. 1066–1069 (1977).

27. I. M. Vitkovitsky, D. Conte, R. D. Ford, W. H. Lupton, "Energy Storage Compression and Switching," Vol. 2, V. Nardi, H. Sahlin, and W. H. Bostick, Eds., Plenum, New York, p. 953 (1983).

28. R. D. Ford and I. M. Vitkovitsky, "Explosively Actuated 100kA Opening Switch for High Voltage Applications," NRL Memo Report 3561 (1977).

29. G. Cooperstein, J. J. Condon, and J. R. Boller, "The Gamble I Pulsed Electron Beam Generator," *J. Vac. Sci. Technology* **10,** 961–964 (1973).

30. S. A. Genkin, Y. D. Kosolev, V. G. Rabotkin, and A. P. Khuzeev, "Spatial Structure of a Volume Discharge Produced by an Electron Beam in Air at Intermediate Pressures," *Sov. J. Plasma Physics* **7,** 327 (1981).

31. S. C. Brown, *Basic Data of Plasma Psychics,* MIT Press, Cambridge, MA, pp. 87–94, (1967).

32. H. Goldstein, Classical Mechanics, Addison-Wesley, Reading, MA, p. 41 (1950).

33. G. Nolting, Naval Surface Weapons Center, private communication (1984).

34. R. O. Hunter, "Electron Beam Controlled Switching," *Proceedings of the International Pulse Power Conference,* Lubock, TX, IEEE Cat. No. 76CH1147-8REG5, IC 8, pp. 1–6 (1976).

35. J. P. O'Laughlin, "PFN Design Interface with E-Beam Sustained Gas Discharge," op. cit., Ref. 34, IIIC5, 1–6.

36. P. Bletzinger, "E-Beam Experiments, Interpretations and Extrapolation," *Proceedings of Workshop of Diffuse Discharge Opening Switches,* Tammarron, CO, Texas Tech Univ. Report, pp. 112–130 (1982).

37. Y. I. Bychkov, D. Yu. Zaroslov, N.V. Karlov, G. P. Kuzmin, G. A. Mesyats, V. V. Osipov, A. M. Prokhorov, and V. A. Telnov, "Pulsed Bulk Discharge with Plasma Cathode in Molecular Gases at High Pressure .I. Externally Maintained Discharge," *Sov. Phys. Tech. Phys.* **28**, 918–920 (1983).

38. P. A. Miller, J. W. Poukey, and T. P. Wright, "Electron Beam Generation in Plasma-Filled Diodes," *Phys. Rev. Letter* **35**, 940–943 (1975).

39. R. Stringfield, R. Schneider, R. D. Genuario, I. Roth, K. Childers, C. Stallings, and D. Dakin, "Plasma Erosion Switches with Imploding Plasma Loads on a Multiterawatt Pulsed Power Generator," *Appl. Phys.* **52**, 1278–1284 (1981).

40. R. A. Meger, R. J. Commisso, G. Cooperstein, and S. A. Goldstein, "Vacuum Inductive Store/Pulse Compression Experiments on a High Power Accelerator Using Plasma Opening Switches," *Appl. Phys. Lett.* **42**, 943–945 (1983); also NRL Memorandum Report 4838 (1982). A theoretical description of the switch (PEOS) is provided by P. F. Ottinger, S. A. Goldstein, and R. A. Meger, *J. Appl. Phys.* **56**, 774 (1984).

41. R. A. Meger, R. J. Commisso, G. Cooperstein, and S. Goldstein, Naval Research Laboratory Report 5037 (1983). Also detailed measurement of current distribution in the PEOS plasma is given by B. V. Weber, R. J. Commisso, R. A. Meger, J. M. Neri, W. F. Oliphant, and P. F. Ottinger, "Current Distribution in a Plasma Erosion Opening Switch," *Appl. Phys. Lett.* **45**, 1043 (1984).

42. S. A. Goldstein and R. Lee, "Ion-Induced Pinch and the Enhancement of Ion Current by Pinched Electron Flow in Relativistic Diodes," *Phys. Rev. Letter* **35**, 1079–1082 (1975).

43. G. Cooperstein, J. R. Boller, D. G. Colombant, A. Drobot, R. A. Meger, W. E. Oliphant, P. F. Ottinger, S. A. Goldstein, D. Mosher, R. J. Barker, F. L. Sandel, S. J. Stephanakis, and F. C. Young, "NRL Light Ion Beam Research Program for Inertial Confinement Fusion," *Laser Interaction and Related Plasma Phenomena,* Vol. 5, H. J. Schwarz, H. Hora, M. Lubin, and B. Yaakobi, eds., Plenum (New York) pp. 105–134 (1981).

44. L. J. Demeter and V. Bailey, "Staged Implosion Plasma Switch for Vacuum Magnetic Energy Stores," *Proceedings of the Second International Conference on Energy Storage, Compression and Switching,* Vol. 2, Venice, Plenum, p. 1023 (1983).

45. D. R. Kania, L. A. Jones, E. L. Zimmerman, L. R. Vesser, and R. J. Trainor, "Experimental Investigation of a Magnetic Gate as a Multimegampere, Vacuum Opening Switch," *Appl. Phys. Lett.* **44**, 741 (1984).

46. P. J. Turchi, "Magnetoacoustic Model for Plasma Flow Switching," *Digest of the Technical Papers of the Fourth IEEE Pulsed Power Conference,* Albuquerque, NM, Cat. No. 83CH1908-3, pp. 342–349 (1983).

47. R. E. Reinovsky, W. L. Baker, Y. G. Chen, J. Holmes, and E. H. Lopez, "Shiva Star Inductive Pulse Compression System," *Digest of the Technical Papers of the Fourth IEEE Pulsed Power Conference,* Albuquerque, NM, Cat. No. 83CH1908-3, pp. 196–201 (1983).

48. P. J. Turchi, private communication (1984).

49. A. Gentilini, Ch. Maisonnier, J. P. Rager, *Comm. Plasma Phys. Controlled Fusion* **5**, 41 (1979).

50. W. H. Lupton, Naval Research Laboratory, private communication (1984).

51. D. D. Hinshelwood, J. R. Boller, R. J. Commisso, G. Cooperstein, R. A. Meger, and J. M. Neri, "Long Conduction Time Plasma Erosion Opening Switch Experiment," *Appl. Phys. Lett.* **49**, 1635–1637 (1986).

52. S. P. Bugaev, B. M. Kovaltchuk, G. A. Mesyats, "High-Power Nanosecond Pulse Generator with a Vacuum Line and Plasma Interrupter," presented at the Sixth International Conference on High Power Particle Beams, Kobe, Japan (1986), unpublished.

10
Solid State Opening Switches

INTRODUCTION

Many solid materials possess potentially important properties for design of high power electrical apparatus. Properties that are of interest to switching are those that allow the solid materials to be controlled, i.e., they can be set to behave either as dielectrics or as conductors. Such materials have already been used for low power switching applications, often for protecting electrical equipment against abnormal operation. The need for reliable and reusable opening switches in very high power pulse generation has only recently led to examination of these materials[1], in terms of their insulating properties, conducting properties, and their transition from the conducting to the non-conducting state, as possible candidates for reusable opening switches.

On the basis of small scale experiments and examination of available data in the literature relating to materials with a strong non-linear dependence of resistivity on external influences, such as heating, magnetic fields, or pressure, a number of commercially available materials were selected as suitable for their further development for pulse power uses[1]. For example, their insulating properties, such as maximum electric field strength and residual conduction, were obtained under more stressful conditions than those associated with their present applications. In addition, samples were evaluated for maximum rate of resistivity transition, so as to provide data connecting the electrical performance with the mechanical stability of these materials, as well as information needed to evaluate what design options might emerge for developing advanced high power pulse generators.

The appearance of materials with non-linear resistivity in commercial quantities and a greater range of electrical characteristics, in particular those whose resistance depends upon current or temperature, gave rise in 1983 to a proposal to exploit solid state devices as opening switches in pulse power applications[1, 2]. (These materials complement the semiconducting materials used primarily as closing switches by activating bulk conduction using intense photon or charged particle beams[3].) The application of non-linear resistivity materials to opening switches is of particular interest to the generation of pulses with power exceeding a 10^{10} watt level using inductive energy storage. Because of the high energy density associated with magnetic fields, inductive storage is often the preferred system, as has been indicated throughout this book, because of the large advantage in compactness of such systems as compared with capacitive storage systems. The use of inductive storage requires, however, a high-current, fast-opening switch to generate the output pulse. This function in high power generators is presently performed by such devices as explosively driven circuit breakers, fuses, and combinations (staging) of these as described in preceeding chapters. Explosively driven switches (Chapter 7), are used in storage systems powered by rotating machinery, for which long conduction times and fast interruptions are essential. Fuses provide substantially faster opening times than explosive switches and are often used in inductive systems for which capacitor banks provide the input current into the inductor. Opening switches, needed in repetitive operation or with very fast opening times, depend on the use of diffuse discharges[4]. Such switches include plasma erosion switches and electron beam controlled ionized gas switches, discussed in Chapter 9. The development of opening switches based on non-linear resistivity materials, only in its infancy at present, offers to replace, ultimately, some of these switches with much more practical devices.

The use of a solid state material with non-linear resistivity, i.e., where the resistance value of the material can change reversibly by several orders of magnitude (e.g., in response to a thermodynamic phase change), provides an alternate approach to the opening switches described in the preceeding chapters. Switches based on the use of such materials exhibit volume conduction changes that, in contrast to arc discharges, do not involve ionization and thus allow current interruption and voltage hold-off to occur simultaneously.

Thermal control of the electrical conductivity of solid materials provides a practical and convenient mechanism allowing these materials to be used for opening switches. The use of such switches would, potentially, include replacement for explosively driven switches and exploding wire fuses in large inductive storage systems. The most important advantage in employing these materials to replace the more conventional opening switches is that they could be reusable, unlike the single-shot explosively driven and fuse switches. Furthermore, when internal heating is used to induce a phase transition in a solid material, control of the current density allows also the conduction time (i.e. the "on" time), to be adjusted from a microsecond range up to a few seconds.

Figure 10-1 shows the non-linear resistance of two commercially produced materials: $BaTiO_3$ ceramic, and carbon filled polymer (CFP). Some of their properties and switching mechanisms are described in Ref. 5 and

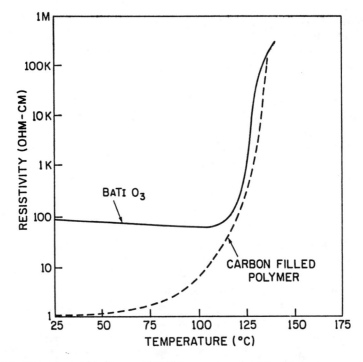

Fig. 10-1. Dependence of resistances of BaTiO$_3$ ceramic and carbon-filled polymer on temperatures. The resistivities of both materials change their behavior above a temperature T_d of about 100°C. Reproduced from Ref. 1. © 1985 IEEE.

6 respectively. These materials exhibit positive temperature coefficients (PTC) above 100°C, associated in BaTiO$_3$ with its Curie temperature, and in the carbon filled polymer with the phase transition of the polymer from its crystalline form to an amorphous state. A change of several orders of magnitude in their resistivity, without the requirement of a recovery time for voltage blocking (i.e., time to hold off the voltage generated by the current drop in the storage inductor), makes these materials suitable for true opening switches, i.e., they change from conductors to non-conductors reversibly. Furthermore, their conductivities can be quite high, so that a relatively low dissipation during conduction would be acceptable in many applications of these materials for switching.

A sufficient flux of photons[7, 8] or charged particles[9] injected into normally high resistivity bulk semiconductors makes these materials conducting. Conduction is restricted to the time interval during which relatively intense radiation is absorbed in the material. As irradiation ceases, conduction decreases, leading to interruption of the current flow. While this effect is equivalent to switch opening, the radiation intensity required to sustain conduction is such that the short conduction times which are available do not exceed the opening time (governed by the carrier decay after radiation cut-off). Thus the bulk semiconducting switches usually are reserved for closing functions, especially where very fast or very accurate closing is needed. So far, switching tests at up to 10^8 watts have been performed[3].

SOLID STATE SWITCHING TRANSITION MECHANISMS

Resistivity increase in a solid can stem from the transition of a superconductor to a normal conductor, from a conductor to a semiconductor, or from a semiconductor to an insulator. Of interest here are materials which undergo such a change reversibly over a convenient range of an input variable, such as temperature, or radiation, which triggers the transition. Such a transition is to be contrasted with the conversion of a fuse from metallic to vapor phase, from which the fuse cannot be reconstituted once its transition to a very high resistivity state has occurred.

A non-linear change in the resistivity of bulk semiconducting materials considered for high power applications is generally attained by a phase transformation that causes a change in structure of the materials on atomic level. The resistivity change is often sufficiently large to satisfy the needs of pulse power switching at power levels yet to be determined. If no electrical or mechanical breakdown occurs, the change is reversible, so that the opening switches made from such materials can be reused. The current density that can be handled by a reversible, reusable switch is, ultimately, limited by the breakdown due to the electric field caused by the ohmic drop across the switch.

The resistivity of material depends on carrier concentration and on the mobility of the carriers. In an arc or in a plasma, the current can be reduced by lowering the carrier concentration, for example, by cooling the ionized channel by sudden contact with a wall. Recombination of free electron carriers with the ions in the background gas (as discussed in detail in Chapter 9) provides another method to increase resistivity. The current in a solid material, on the other hand, can be lowered by reducing the mobility of the carriers, for example, as a result of heating, which alters the electronic structure of the conducting material. The latter switching mode makes it possible to increase the voltage across the device at a rate determined by the temperature rise, i.e., the power input into the material. This generally can be done faster than the removal of the carriers from high current arcs.

There are a number of solid material configurations that are amenable to switching, each having unique features. First of these types is a planar *p-n* junction which concentrates the switching losses in a thin, almost two dimensional zone. Its application to high power switching has not been successful up to this time because it requires use of a heat sink to be applied to a large area; even then the low thermal mass of the junction can easily lead to unacceptably high spot temperatures during rapid pulse switching, which damages the junction permanently.

Another configuration, more likely to be exploited for switching, is low resistance polycrystalline material with grain boundary barriers comprising, in effect, an interconnected three dimensional network of series-parallel switching junctions interspersed with "inactive" carrier material. During switching, this carrier material serves as a convenient sink for the heat generated in the switching regions. This reduces the rate of temperature rise in the barrier and lowers the temperature of any hot spots that may form. Furthermore, the shorting of a single barrier has no noticeable effect in the three dimensional grain boundary network, effectively providing a longer lifetime for such devices. After the grain boundaries are switched to a high resistance, they must hold off the applied voltage since there is no significant voltage drop across the lower resistivity grain bulk regions. Such partial utilization of the material increases the switch volume relative to that needed by a true bulk (homogeneous) switch. This extra volume tends to contribute to the resistance of the device during conduction, requiring an even larger volume to maintain a given low resistance in the "on" state. As stated, the advantage of the added volume is that it acts as a sink for the heat generated in the boundary layers.

Finally, the most promising switching material may be bulk material with carrier mobility changing throughout all its volume. Such materials can provide, in principle, the highest volumetric efficiency since they utilize the entire volume of material uniformly on both microscopic and macroscopic scales. As in the foregoing configurations, resistivity change can be triggered by an external input such as a change in temperature, pressure, or magnetic field.

Thermally-Induced Transitions

Some special features of the more interesting grain boundary materials, as well as those of true bulk switching materials, are considered next. Both inorganic and organic polycrystalline materials have been used as switches, but also some single crystal configurations appear to have very promising properties for this application. The following discussion of these materials is based on the categories in Ref. 1.

The *inorganic polycrystalline* systems applicable to switching are multiphase materials, where often a transition in one phase causes a conductivity change in another phase. In a switch based on grain boundary materials, the grain bulk holds and conditions the grain boundary phase. The grain is also a source of current carriers when the grain boundary phase is in its low resistance condition. The division of functions facilitates the optimization of the different phases during device manufacture, since the grain boundary phase responds to changes in composition and processing in a different way from the grain bulk phase. For instance, polycrystalline barium titanate can be utilized as a grain boundary switching device even though it is generally used as an insulating dielectric[17]. It can be made semiconducting with a resistivity as low as 10 Ω-cm by dissolving in it a small amount (<0.25%) of a donor impurity (i.e. lantanum or niobium). After it has been sintered to a semiconductor, a lower temperature annealing treatment is applied to diffuse oxygen into the grain boundaries, creating acceptor sites. The much lower oxygen diffusivity of the donor-doped grain bulk prevents the creation of acceptor sites in the grain bulk. Grain boundary acceptor sites create a space charge that impedes the flow of conduction electrons, thereby raising the resistance at the grain boundaries significantly.

Below 125°C, barium titanate contains poled domains and charges in the domain walls. Grain boundaries are usually coincident with the walls of domains, and below 125°C barium titanate has therefore charged grain boundaries. Half of these regions contain negative charges that neutralize the space charge caused by the oxygen-induced acceptor states and permit easy flow of conduction electrons across the grain boundaries. At the 125°C transition temperature of $BaTiO_3$, the domain walls and charges disappear. The grain boundary resistance goes up and the effective resistivity of the ceramic increases by three or more orders of magnitude, as demonstrated by the temperature dependence of the resistivity shown in Fig. 10-1. This is called the *positive temperature coefficient resistivity,* or *PTCR,* effect.

Examples of *organic polycrystalline* switches are polymer-carbon composites that also increase their resistance upon heating. They are made of polyethelene filled with carbon particles. The carbon constitutes about 50% of the composite and links itself into chains that lead to relatively low resistivity (about 1 Ω-cm) at ambient temperature. Heating of the device above 85°C causes a phase change in the polymer that is accompanied by a volume increase of about 2%. This breaks the conducting carbon chains. Thin insulating regions then form between the carbon particles, increasing the resistance of the device significantly. The effect of temperature on the resistivity of one composite material is shown in Fig. 10-1. Upon cooling, recrystallization of the polymer and relinking of the chains occur and the device recovers to approximately 90% of its initial conductivity. Test devices have been cycled 10 times at approximately 1 minute intervals at the 90%

recovery state, without further degradation. When left inoperative, relinking of carbon chains continues, with recovery to the initial state occurring over a 24 hour period. (These devices exhibit a resistance drop when external pressure is applied, since this brings the carbon particles into more intimate contact).

In contrast to the types of inorganic and organic materials, which have found commercial use at low power (polymer-carbon composites are used to protect 50 volt circuits from overheating or from current surges[10]), the true bulk-switching materials appear with useful properties only when in single-crystal configuration. Some materials that exhibit true bulk switching are the following. Single crystals of chromium-doped vanadium sesquioxide shows a resistivity increase of 2 orders of magnitude at about 80°C. The change is smaller in commercially available polycrystalline devices; the degradation of performance results from internal stresses in the polycrystalline compacts and from the structural disorder near the grain boundaries. Another material is europium oxide, which has a metallic-to-semiconductor transition at about 50°K. In single crystals, its very low resistivity (typically 0.005 Ω-cm) increases due to heating by as much as 5 orders of magnitude. In addition, the resistivity transition is shifted to higher temperatures in the presence of a magnetic field[11]. This property of europium oxide could be particularly interesting to high power switching applications, since it can be used to maintain a uniform transition in a large volume switch by immersion in the magnetic field generated by the current controlled by the switch. Transition uniformity would be attained through the resulting feedback loop, where the onset of switching (high resistivity) in one segment of the switch leads to a reduction of the current, lowers the field, and induces switching in other parts of the material which had remained slightly cooler.

Transitions Induced by Magnetic Fields

Superconducting materials also represent bulk switching material. They attain their normal resistivity as a reuslt of heating or of an increase in the magnetic field (e.g., increasing the current above the critical limit). The abrupt transition to finite resistance has already been exploited in development of superconducting switches[12, 13]. Their main limitations are their relatively low operating voltage and low resistivity in the "off" state. In addition, in practice it becomes quite cumbersome to control the heat flow through the lead-in wires, making switch construction very complex. Very preliminary research[14] suggests that it may be possible to build superconductors with "granular" microstructure having resistivity as high as 0.1 Ω-cm. If such materials can be developed in relatively large quantities, switching using superconductors may become much more practical.

Finally, an increase of resistivity triggered exclusively by a magnetic field is found in magnetoresistive materials. For example, indium antimonide exhibits a significant resistivity increase, associated with its extraordinarily high carrier mobility, when immersed in a magnetic field. A tranverse magnetic field causes the current to be deflected away from the opposite electrodes. This phenomenon has been evaluated in a power switch, where this property is utilized in a doughnut-shaped "Corbino" disc[15] inserted into a very strong magnetic field, about 50 kilogauss. The carriers follow a spiral path, increasing the effective resistance by two orders of magnitude. Current densities can be as high as 3 kA/cm^2 and a capability to withstand a voltage of 1.5 kV/cm can be attained. This material appears to be a prime candidate for a high power switch in systems that generate large magnetic fields. Indium antimonide can be utilized at room temperature. No significant grain boundary resistance has been observed in its polycrystalline form. It melts at 535°C, crystals are relatively easy to grow, and boules in excess of 5 cm in diameter can be fabricated. Casting of polycrystalline ingots should permit the preparation of even larger pieces.

There are a number of materials that exhibit non-linear negative voltage current (*NTCR,* or *negative temperature coefficient resistor*) characteristics. These materials are not useful directly for opening switches, but may be applied in subsidiary roles in inductive storage systems, for example, as crowbar elements and overvoltage protection devices. In general, though, materials that exhibit a resistivity drop with increasing temperature respond more slowly to an input trigger than PTCR (positive temperature coefficient resistor) devices because they require a flow of heat from external sources to change the temperature throughout their volume. This is a slower process than Joule heating and harder to implement. One of the NTCR devices is vanadium oxide (VO_2), which drops in resistivity by 3 orders of magnitude when heated above 65°C. Similarly, vanadium doped Ti_2O_3 looses resistance when the temperature of 150°C is exceeded. Chalcogenide glasses[16] can be used for resistivity switching, induced by current triggering. It is thought that in these materials the current causes localized heating which generates low resistivity "filaments."

Transitions Induced by Electric Fields

One family of switching materials, polycrystalline zinc oxides, change their resistivity in response to changes in electric field. In these varistors, the grain bulk acts as a carrier for an "active" grain boundary phase and it is the latter that shows a strong drop in resistivity when a critical electric field is exceeded. The number of the grain boundaries between the electrodes determines the device's switching level, typically at a fraction of 1 kV/cm.

It has been mentioned above that a limit on the current density in the switch is imposed by the breakdown of the solid material by the applied electric field. I. Ueda et al.[17] have reviewed the studies of breakdown in single crystal and polycrystalline $BaTiO_3$ *high* resistivity samples and report on their experiments with dc and pulsed field breakdown in this material. Their results, such as those seen in Fig. 10-2, show that electric fields in the range of 100 kV/cm can be sustained below and above the Currie temperature. The pulse duration in these experiments varied from 100 μs down to 4 μs.

The intrinsic breakdown fields, under pulsed conditions, in *low* resistivity materials and in materials where the resistivity changes from initially low to final high values have not been evaluated under pulsed conditions. Other limitations and difficulties in determining such fields are the subject of the next section, "Limitations on Switching."

Photo-Induced Transitions

The photoconducting materials used for switching[18] behave as true bulk materials. This category of materials requires injection of energy to provide conduction. Turn-off of the energy sources reverts these materials to the non-conducting state. Though pulsed sources such as lasers[7] or electron beams[9] can be employed to obtain sufficient irradiation intensity, their short durations limit the conduction times; they cannot compete with such materials as barium titinate or CFP for long conduction times. Thus, these materials are not normally associated with opening switches. However, their switch closing time can be extremely short—in the nanosecond and even the picosecond regime[18, 19]. Switches made from photoconducting materials should not be confused with light-activated silicon switches (LASS)[20]; these two types of switches differ from each other in at least two respects. LASS uses a four layer thyristor structure with three junctions, with one of the junctions serving to hold-off the applied voltage before switching. The photoconducting power switch (called PCPS in Ref. 3) distributes the applied voltage across its entire length, in a manner similar to that in non-linear resistivity materials. Furthermore, the LASS device uses the photoconductive effect to change only the conductivity of its gate region. The PCPS, on the other hand, is uniformly illuminated to change the conductivity of its entire volume simultaneously[3, 8]. This feature is also responsible for the faster switching capability of the PCPS devices. The breakdown strength of photoconducting silicon is predicted to be in the 100 kV/cm range[3], similar to that for ceramics shown in Fig. 10-2.

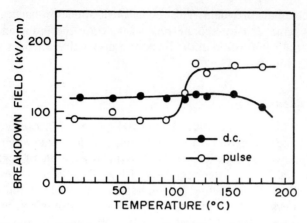

Fig. 10-2. Breakdown field of commercial grade BaTiO₃ under dc and pulsed conditions. Reproduced from Ref. 17.

LIMITATIONS ON SWITCHING

PTCR Switches

The resistivity transition in a PTCR switch is caused by a temperature increase and the maximum permissible rate of the resistance increase depends upon the allowable rate of temperature change. The latter is limited by temperature gradients that inevitably develop with fast heating rates and that can attain magnitudes which cause failure from thermal shock. Thermal gradients can arise from resistivity inhomogenuities that exist in all real devices. With rapid resistive heating, they give rise to potentially damaging stresses.

There are also other sources of thermal gradients in PTCR's. These occur because the electrodes comprise a thermal mass that retards heating near them just long enough to allow the center of the device to increase in resistivity before the outer regions do. The higher resistivity regions are subjected to increased power dissipation densities, which can then lead to localized thermal runaway. The latter effect occurs when temperatures exceed 250°C, beyond which the resistivity of a PTCR device starts to decrease again with temperature. In addition, the voltage distribution becomes distorted by the resistivity gradients. Even in the most homogeneous devices, excessive electric fields can then occur, resulting in Zener breakdown.

Finally, fast pulse risetimes make it important to minimize the inductance of the switching arrangement. This requirement applies in particular to the PTCR packaging and interconnecting leads, since the inductance of the PTCR material itself, often in the form of a disc with electrodes on its flat surfaces, is minimal.

PCPS Switches

G. Mourou et al. discuss in their chapter "High Power Picosecond Switching in Bulk Semiconductors" (Ref. 19) optically controlled switching in relatively low power applications, such as synchronization of high voltage pulses in inertial fusion systems, control of Pockels cells, and streak camera operation. The authors define the limitations imposed on the switching waveforms. Because the "on" state of the switch is obtained by creation of carriers by the externally injected photons, the optical input determines to a great degree the switching waveforms. In most cases, the absorption length in the semiconductor is much less than the length of the input optical pulse as given by the product of the velocity of light in the material and pulse duration, so that the carrier build-up time is determined by the optical pulse waveform. Thus, the risetime and jitter of the electrical pulse will be smaller than the pulse duration.

Other characteristics of the electrical pulse, generated by the application of a voltage across the illuminated bulk semiconductor, are determined by the following three conditions. First, carrier recombination establishes the upper limit on conduction time when the switch is irradiated by short pulse (with a duration less than the carrier lifetime, which is, for example, 10^4 ns, 50 ns and less than 1 ns for the intrinsic Si, Au:Si, and Cr:GaAs semiconductors, respectively). Second, if the carrier lifetime is long, the depletion of energy from the voltage source must be considered a potential limit on the pulse duration. (This can be a significant limitation when optically controlled switches are used in conjunction with charged transmission lines for precise pulse shaping.) Third, when the recombination time is short relative to the optical pulse duration, the electrical output waveform will approximate the optical pulse profile so that the latter pulse determines the output pulse waveform. These three factors must be considered together with the electrical circuit to be switched to determine whether PCPS can be used effectively for current interruption.

In the generation of long pulses, as well as in repetitive applications, another factor that limits the PCPS is the Joule heating of the switching material generated by the electric field. Such heating leads to thermally generated carriers, culminating in the thermal runaway that destroys the switching element.

The optical pulse energy required to control PCPS can become significant at high power levels. Mourou et al.[19] estimate the required optical energy, W, by considering the gap (of length ℓ) resistance, R_0, after the excitation (and before recombination) to be (in mks units)

$$R_0 = h\nu\ell\, V_0/2veW \quad \text{(ohms)} \tag{10-1}$$

where $h\nu$ is the energy of the incident photon, V_0 is the gap voltage, v is the carrier saturation velocity (of about 10^7 cm/s) and e is the electron charge. In the high field (> 10 kV/cm) approximation, the electron and hole currents are equal. Assuming, that 95% of the gap voltage is to be switched (i.e., a pulse voltage of 0.95 $V_0/2$ corresponds to a switch resistance $R_0 = 0.1Z$, where Z is the source impedance), then

$$W\ (95\%) = 5V_0\ell\, h\nu/Zve \quad \text{(joules)} \tag{10-2}$$

For $\ell = 0.6$ cm, $Z = 50\ \Omega$ and $V_0 = 8$ kV, the optical energy turns out to be 40 μJ. As pointed out in Ref. 19 by Mourou et al., in practice this is an underestimate by a factor of less then 10.

The amplitude of the current pulse in the above example is about 150 A, i.e., the switch operates at about the 1 MW level. The application of the photoconducting power switch (PCPS) at much higher power levels ($\lesssim 10^8$W) has been evaluated by W. C. Nunnaly et al.[3] using 1 kΩ-cm n-type silicon, 2.5 cm long bars with a 0.5 cm $\times$ 0.5 cm cross-section. Photoconductivity was induced by 1.06 μm photons from a Q-switched laser (varying in pulse energy from 0.03 to 3 J). The active silicon volume was limited by the transparency of silicon. The laser light penetrated to a depth of not more than 1 mm. From those experiments, 20 kA per cm of length (transverse to current direction) is projected. As already noted, the hold-off voltage should be in 100 kV/cm range, with tests indicating a 40 kV/cm hold-off. Large specific heat and excellent thermal conductivity make this device promising for high power switching. Repeated operation shows no degradation. The major constraint arises from the carrier density in silicon, limited by Auger recombination, free-carrier absorption, and local thermal runaway considerations to about 10^{18} cm^{-3}, corresponding to a maximum surface optical density of about 10 mJ/cm^2. With a 1 mm optical depth, the current limit is 20 kA/cm. These limits translate[3] into a switched energy gain of about 10^5. Since the efficiency of 1.06 μm laser is about 1%, the overall gain is 10^3.

PTCR SWITCH DESIGN GUIDELINES

Scaling

Two characteristics that determine how suitable a switch is for a given application are its opening time and its efficiency in energy transfer. These characteristics depend to some degree on external factors such as the current to be carried by the switch and on the configuration and the values of the associated circuit components. To provide general expressions for scaling of internally heated PTCR switches, the conduction time before the start of switching, as well as limiting (lower) values for opening time and for efficiency relative to that of a fuse opening switch, have been derived. These expressions provide comparison with the energy losses, conduction times, and opening times of the more familiar fuse opening switches discussed in Chapter 8.

The opening time of the ceramic $BaTiO_3$, which has switching characteristics determined by its narrow temperature range (of about 20°C), is considered first. The narrow switching temperature makes this material adaptable to triggerable operation, i.e. for use in circuits where the conductivity change can be induced by a relatively small energy increment. In addition, the narrow temperature range allows the material to return to its initial conducting state with only a relatively small amount of heat removal. (In contrast, carbon-filled polymer, CFP, must be cooled almost to room temperature to recover its initial resistivity because of its substantial increase of resistance at temperatures well below its transition temperature. This can be seen by comparing the two curves in Fig. 10–1). Thus $BaTiO_3$ can potentially be used in burst output pulsers (based on circuits described in 7–13B and in Ref. 21) with pulse-to-pulse separation more closely spaced because less heat removal is necessary. On the other hand, the cold resistivity of $BaTiO_3$ is substantially higher (10–100 Ω-cm) than CFP, so that its applications may be limited by physical size to lower currents than those of the CFP switch.

To determine the time to the on-set of switching, the resistivity curves in Fig. 10–1 are divided into temperature-dependent regions, which are now considered. The first region (below temperature $T_d \simeq 100$°C), has a relatively weak linear temperature dependence of resistivity (with a slightly negative slope for $BaTiO_3$). The second domain above T_d has a resistance that has an approximately exponential increase with temperature. The delay of the switching on-set, t_d (associated with the linear region), is thus inversely proportional to the absorption of power, P, (i.e., I^2R), that heats the PTCR switch up to temperature, T_d:

$$t_d = \left(\int_{T_0}^{T_d} MC_v \, dT \right) / P \qquad (10\text{-}3)$$

Here T_0 is the ambient temperature, M is the mass of the switching material, and C_v its specific heat capacity (0.5 J/g°C). Assuming that in pulsed operation, no significant heat is lost to the environment by the switch, the delay time is

$$t_d = M C_v (T_d - T_0)/P \qquad (10\text{-}4)$$

For a 100°C PTCR switch that is initially at $T_0 = 25°C$, the delay (in seconds) is $t_d = 37.5 \, M/P$. As an example, $BaTiO_3$ samples weighing 20 grams can absorb a power density of 0.75 kW/g without breakdown. Thus, the minimum delay time is $t_d = 0.05$ s. If the Joule heating continues, then the resistivity begins to rise at a much higher rate. To determine the switching interval itself, the procedure following the derivation by A. Robson[22] and outlined in Ref. 1 can be used. The assumption of zero heat loss leads to a simple relation, showing that the conducting material increases in temperature, T, at a rate proportional to the electric power density input:

$$\frac{dT}{dt} = \frac{\rho j^2}{C_v} \qquad (10\text{-}5)$$

where C_V is again the specific heat capacity, ρ is the resistivity of the material and j is current density. Approximating the resistance characteristic shown in Fig. 10-1 by an exponential, $\rho(T) = \rho_0 e^{\alpha T}$ the solution of Eq. 10-5 becomes

$$e^{-\alpha T} = 1 - \frac{\alpha \rho_0 j^2 t}{C_v}, \qquad (10\text{-}6)$$

where t is the current fall time interval. If the initial resistance increases by a factor γ, such that $\rho = \gamma \rho_0$, then

$$\frac{1}{\gamma} = 1 - \frac{\alpha \rho_0 j^2 t}{C_v} \qquad (10\text{-}7)$$

and

$$t \simeq C_v / \alpha \rho_0 j^2 \quad \text{for } \gamma \gg 1 \qquad (10\text{-}8)$$

Equation 10–8 can be written in terms of the applied field, $E = \gamma j \rho_0$,

$$t = C_v \rho_0 \gamma^2 / \alpha E^2 \qquad (10\text{–}9)$$

One limiting factor in Eq. 10–9 is the maximum hold-off field of the switch, E_m. To obtain minimum current interruption time, t_{min}, the maximum value of j that can be passed by the non-linear material is used (i.e., $j = E_m / \gamma \rho$). The value of t_{min} then is

$$t_{min} = C_v \rho_0 \, \gamma^2 / \alpha E_m^2 \qquad (10\text{–}10)$$

This relationship indicates quantitatively the importance of the withstand (breakdown) voltage of the non-linear materials in their ability to switch rapidly. Additionally, other effects, such as thermal shock, may limit performance further.

Unlike copper wire fuses, for which voltage withstand capability versus conduction time is well documented, PTCR switching devices have been measured only under a very limited set of pulsed power conditions. Inductively generated fields of 3 kV/cm to 15 kV/cm have been observed (with details given in a later section). The breakdown data from Ref. 17, quoted earlier, suggest that better PTCR materials may be developed to withstand 50–100 kV/cm inductively generated fields. When evaluating this, the physical limitations of material configuration must be considered: hold-off voltages above 3 kV/cm were obtained in thin wafers (≤ 1 mm) where thermal equilibrium is rapidly attained[1]. It is pointed out in Ref. 1 that in thicker samples, non-uniform heat distribution may produce localized electric fields that exceed the breakdown strength at lower average fields.

Comparison with Fuses

In a pulsed power system, where the switching energy is derived from the energy storage circuit, the switching efficiency of the non-linear resistance material is of interest. Note again from Fig. 10–1 that the resistivity of cold CFP is about two orders of magnitude lower than that of $BaTiO_3$, so that the low voltage drop (during conduction) allows the CFP material to be used in a manner similar to the exploding fuses employed in inductive storage circuits. A comparison of CFP devices and fuses in terms of energy required for switching can then be made more meaningful than would be the case for barium titanate in terms of switch efficiency. Denoting quantities referring to fuses by the symbol F and those to CFP by the symbol C, relations for the energy deposited into the switch at the time that switching occurs can be defined.

For fuse wire:
$$W_F = \int_{T_0}^{T_F} M_F C_F \, dT + w_F \qquad (10\text{--}11)$$

(and the value of W_F is equal to the $w_0$hs given by Eq. 8-28.)

For carbon-filled polymer:
$$W_C = \int_{T_0}^{T_c} M_C C_C \, dT + w_C \qquad (10\text{--}12)$$

where T_0 is room temperature, T_F is the vaporization temperature of copper in the fuse, and T_C is the transition temperature ($\sim 120°C$) of the polymer with dispersed carbon from its low to its high resistance state. M represents the masses of the switches; C, their specific heat capacities; and w, the energies associated with latent heats of tansition.

In order to compare the expenditure in switching energy of the two materials, it is assumed that they have the *same resistance* during conduction (i.e., cold resistance). It is also assumed that the *final inductive voltage* to appear across the switching elements is V. Thus, for respective switch lengths L_F and L_C, the voltage is limited by the breakdown field, E, of the material, so that $V = L_F E_F = L_C E_C$. The ratio of the two switching energies from Eqs. 10–11 and 10–12 is

$$\frac{W_C}{W_F} = \frac{M_C \int_{T_0}^{T_C} C_c \, dT}{M_F \int_{T_0}^{T_F} C_F \, dT} = \frac{M_C}{M_F} Y \qquad (10\text{--}13)$$

where $W - w \simeq W$ and Y is the ratio of the integrals. Denoting densities of the materials by δ_F and δ_C, and cross-sectional areas by A_F and A_C, Eq. 10-13 becomes

$$\frac{W_C}{W_F} = \frac{\delta_C A_C L_C}{\delta_F A_F L_F} Y = \left(\frac{\delta_C}{\delta_F}\right)\left(\frac{A_C}{A_F}\right)\left(\frac{E_F}{E_C}\right)^2 Y \qquad (10\text{--}14)$$

Denoting the respective resistivities ρ_F and ρ_C, the requirement of equal initial switch resistance, $\rho_F L_F / A_F = \rho_C L_C / A_C$, yields

$$\frac{W_C}{W_F} = \frac{\delta_C \rho_C L_C E_F}{\delta_F \rho_F L_F E_C} Y = \left(\frac{\delta_C}{\delta_F}\right)\left(\frac{\rho_C}{\rho_F}\right)\left(\frac{E_F}{E_C}\right)^2 Y \qquad (10\text{--}15)$$

For copper fuses, $\delta_C/\delta_F \simeq 0.1$ and, if $\rho_C = 1 \ \Omega\text{-cm}$, $\rho_C/\rho_F \simeq 3 \times 10^5$. Values of E_F are well documented (and discussed in Chapter 8); they vary from 20 kV/cm at microsecond conduction times to less than 0.2 kV/cm for millisecond times.

Comparable data is not available for CFP materials, but have been obtained experimentally for test samples at millisecond conduction times. Test values greater than 3 kV/cm have been observed for thin (< 1 mm) test samples[1]. Thus, for the purposes of this analysis, E_F/E_C is approximately 0.1 (as can be seen from Fig. 8–13 for a conduction time around one millisecond), and $W_C/W_F \simeq 300 \ Y$. To obtain the approximate value of Y, consider that the conduction in CFP is by graphite only, so that the temperature integral of C_C changes linearly between room temperature and the transition temperature of $\simeq 100°C$. The integral of C_F for copper (for temperatures ranging from room to vaporization values) is about 30 times larger. Thus, for millisecond conduction times, $W_C/W_F \simeq 10$, and approaches unity for $E_C \simeq 30$ kV/cm.

These equations suggest that CFP switches that are reuseable compare relatively well with one-shot exploding wire fuses. The data are only preliminary; the performance of CFP switches relative to that of fuses would change as the conduction times are changed. Experimental verification of the scaling is also reported in Ref. 1 and is described below, with results listed in Table 10–1.

MEASUREMENTS OF ELECTRCAL PROPERTIES OF SWITCHING MATERIALS

PTCR switching devices available for pulse power testing are commercial components rated to operate at tens to hundreds of volts and 1 to 10 amperes[23]. Figs. 10–3 and 10–4 are photographs of PTCR and CFP materials respectively. The barium titanate PTCR has electrodes coated at the ends. The CFP is shown encapsulated with its electrodes as a thin wafer. According to the manufacturer's specifications, the CFP devices are able to interrupt or limit the rated current and rated hold-off voltage for long times, until the user interrupts the power. Should the device be operated at voltages or currents in excess of specifications, thermal runaway or thermal shock leading to component destruction would occur.

These same components, when used in a pulsed power configuration, may be operated at many times their nominal rating, since at the time Joule heating approaches thermal runaway conditions, the pulsed power source will have been depleted and no additional heating occurs. This is evident from Fig. 10–5, which shows the reduction of the time to switch with increased switch current. Many pulsed power applications are concerned with

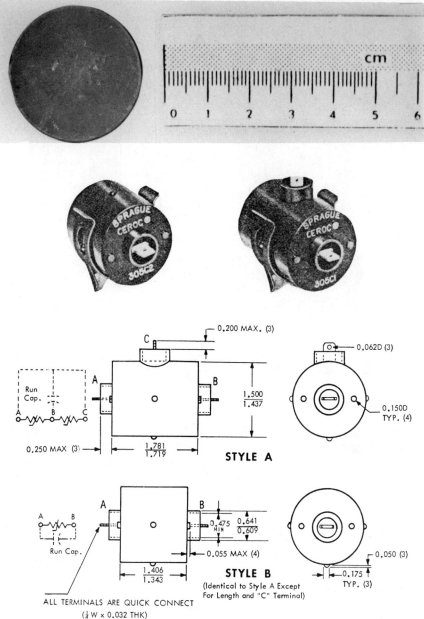

Fig. 10–3. BaTiO$_3$ PTCR switching element and packaging for quick plug-in. Dimensions are given in English units (inches). Reproduced from Ref. 23.

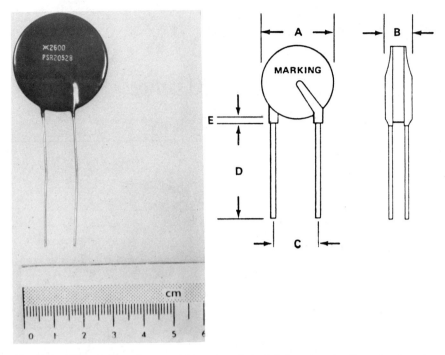

Fig. 10–4. CFP switching element with metric scale included to show its dimensional specifications with diagram at right. Reproduced from Ref. 10.

even shorter times than shown in the figure, indicating much higher current capability. Tests of components under pulsed power conditions have usually excluded thermal runaway intentionally[1] in order to identify the upper limits to the components' performance in terms of voltage hold-off, current capacity, and switching time.

Tests of BaTiO₃-Based PTCR Devices

Ceramic PTCR's are used widely in commercial applications as motor-starting devices. The Sprague Company's Type 305-C device[23] is available in easy to plug-in packages, shown in Fig. 10–3. Also shown is the barium titanate disc used in such packages, which is available with cold resistance ranging from 100 Ω to 25 Ω. Since it is likely that lower resistivity devices will be preferable in high power applications, most tests have been performed at the Naval Research Laboratory using Keystone Carbon part RL5405-3.0-120-20-PTO; it has a nominal resistance of 2.5 ± 0.5 Ω and a

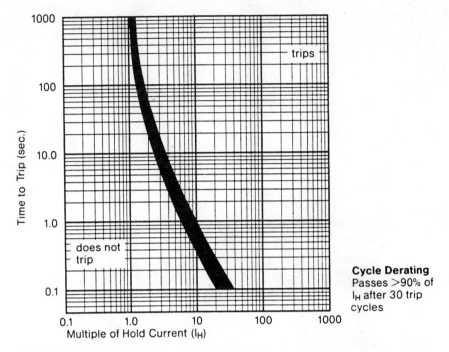

Fig. 10-5. Dependence of the time at which switching occurs (time to trip) on the current. The current value is normalized to the maximum current at which no switching occurs. Figure reproduced from Ref. 10.

maximum voltage rating (for a steady state) of 20 V[1]. It measures 1.4 cm diameter by 0.12 cm thick and has a density of 5.5 gm/cm³. Heat absorption of the material is 37.5 J/gm in the operating range from room temperatures to switching temperature. Samples have been tested in the circuit shown in Fig. 10-6.

When a 20 A dc power supply was used to heat the PTCR switch, it was observed that for a conduction time of 8 seconds the shut-off interval (i.e., the current fall time) of the device was 1.5 seconds. The switching time could be reduced to approximately 1 ms without affecting conduction time when a 1200 μF capacitor charged to 40 V was used as a triggering current source and discharged into the switch under test at the onset of the switching action. Switching efficiency (based on inductor energy transferred to load vs. trigger source energy) was observed to be 95% under these conditions. The operating performance for repeated cycling was not degraded by the application of the 400 V trigger pulse, provided the circuit limited the thermal absorption of the PTCR to 37.5 J/gm.

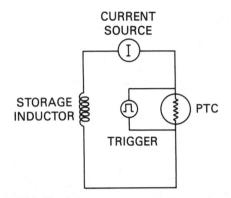

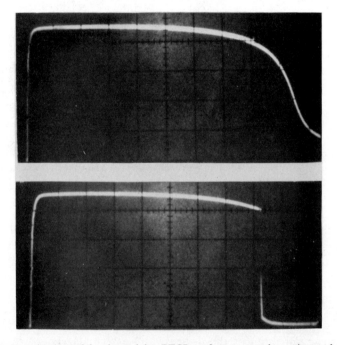

Fig. 10–6. Test circuit used for determining PTCR performance under various pulsed power conditions and an example of the switching characteristics. The top trace shows the effect of ohmic heating by the source current and the bottom trace illustrates switching controlled by the external trigger source. Each division in the grids represents 0.5s. © 1985 IEEE.

Reference 1 describes further observations of component response under pulsed power conditions, when tests have been performed using a 20 mF capacitor bank charged to 200 V connected in series with a pulse shaping inductor, and a current-clamping diode. Peak currents of more than 40 A having pulse duration (time to switching) greater than 80 ms were possible with this configuration. Additionally, a 3.6 mF capacitor charged to 140 V and discharged directly into the switching device (in a manner shown in Fig. 10–6) at onset of switching, was used for triggering purposes. It was determined that the device could be tested repeatedly with conduction time of 80 ms and a triggered switching time of 1 ms without degradation in performance. Switching efficiency of 90% was measured.

Components operating in parallel were observed to share the current, with current fall times similar to that of a single component. However, in order to operate PTCR's in series, it was necessary that an external load-sharing device be added to each PTCR.

During these tests, it was found that the maximum voltage withstand capability for these components was 3 kV/cm. At higher fields, failure occurred in the form of surface cracking. This type of failure appeared to result from non-uniform heating in localized areas, rather than from voltage-induced surface flashover, or breakdown of the material in a manner described in Ref. 17.

Measurements of Carbon-Filled Polymer Device Performance

Carbon-filled polymer material has a resistivity substantially lower[6] than that of commercially available $BaTiO_3$. It is available as a current limiter device under the trade name Polyswitch[10]. For most tests, Raychem production part PSR20528 has been used. It is shown in a plastic package in Fig. 10–4. Its resistivity is in the 1 Ω-cm range with steady state voltage and current ratings of 15 V and 9 A, respectively. These components have been tested with inductively stored pulse currents up to 10kA, at short times, and with conduction times at lower currents to 500 ms. This has been done repeatedly, without damage to the parts.

Reference 6 indicates that small signal resistance ratios R/R_0 on the order of 10^5 can be attained, where R is the maximum switch resistance at $\sim 150°C$ and R_0 is the room temperature resistance. However, in a pulse current test in which 600 A of current were switched off after about 500 ms, resistance ratios of only 10^2 were measured[1], i.e., the resistance of the CFP went from 9 mΩ to about 1 Ω. This gave adequate load current switching since the load resistor was 0.125 Ω. The voltage across this load went up to a peak of 75 V, which did not damage the CFP. In this application, it therefore operated as a resettable device.

To obtain test data on the CFP switches under conditions that can be scaled for high power inductive storage designs, they also have been operated as fuses (i.e., as one-shot devices) and tests have been performed to compare their behavior with that of a copper fuse exploded in air[2]. An energy storage inductor was connected to a 12,000 μF capacitor to provide a sinusoidal current pulse rising to a peak current of approximately 2000 A after 2.5 ms. A single CFP switch measuring 30.5 mm diameter by 0.4 mm thick (0.292 cm^3) was observed to switch after a conduction time of 2.5 ms, with a switching time of less than 100 μs. The inductive voltage produced was 600 volts, corresponding to a breakdown voltage field of 15,000 V/cm. No attempt has been made to trigger the switch for faster operation.

With the same pulse circuit being used, an array of copper wires (of 0.25 mm diameter) was selected to obtain switching characteristics similar to those of the CFP. Peak voltages from the copper wire fuses varied from shot to shot, but were nominally the same as those for the CFP switch. Fig. 10-7 shows current and voltage traces for each device and Table 10-1 provides the measured values for assessing polyswitch performance and scaling to higher power application[1]. Maximum values are shown, with breakdown (or restrike) occuring in both examples, effectively as a limit on the generation of higher inductive voltages. Although switching energy ratios of a factor of 10 favoring the wire fuse are predicted from Eq. 10-15, the CFP switch actually performed at least as well as a wire fuse. This result is attributed to the following. First, at the millisecond time scale of this test,

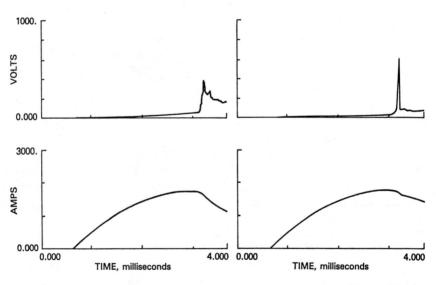

Fig. 10-7. Comparison of fuse and CFP switching peformance. At left, voltage and current waveforms resulting from fuse switching are shown. At right, corresponding waveforms for the CFP are given. © 1985 IEEE.

Table 10–1. Comparison of Dimensions and Electrical Characteristics for CFP and Copper Fuse.

	Conduction Time (ms)	Switching Time (ms)	I_{max} (kA)	R_0 (mΩ)	Area (cm²)	Length (cm)	$\int VIdt$ (J)	Stored Energy (J)	V_{max} (V)
CFP	2.5	0.2	2	9.8	3.050	0.04	60	400	600 ± 20
Copper fuse*	2.5	0.2	2	4.23	0.002	5.0	75	400	600 ± 110

(*Copper wire fuse consists of 4 No. 30 wires 5 cm long connected in parallel.)
© 1985 IEEE.

measured values of E_F/E_C for the conditions of the experiment have been approximately 0.25 instead of 0.1. (With shorter conduction times, E_F increases rapidly, shifting in favor of wire fuses the values predicted by the Eq. (10–15). Tests needed to provide short conduction time values for E_C have not been carried out). Then, additionally, when reusability of the CFP switch is required or desirable, it is necessary to limit E_C to values significantly less than that which destroys the device, and the ratio E_F/E_C should be limited to about 0.1, as the scaling analysis results discussed above indicate.

It should be also noted that the results shown in Fig. 10–7 indicate CFP switching is "smoother", i.e., has a higher rate of voltage rise than the fuse. This initial result suggests the switching reproducibility of CFP devices may exceed that presently available with fuses.

COMMENTS

One significant result of the tests described in Ref. 1 is the demonstration of the difference between volume conduction and channel discharge switching. In circuit breakers and fuses, where an arc must be cooled to stop conduction, the occurrence of a subsequent voltage pulse (as from a second stage switch) must be delayed if the arc is not to divert the current from the load. Such delays, called "recovery" times, are discussed in Chapters 8 and 9. The severe consequences affecting the efficiency of pulse-forming in very high power applications that result from the need of prolonged recovery are discussed in Ref. 24. Switches that depend upon volume conduction can avoid the need for the recovery period and thus eliminate one major source of switch inefficiency.

Projected further into the future, as the tentative results indicate, are BaTiO$_3$ PTCR's with a resistivity in the 10 Ω-cm range, making them usable for repetitive switching in hold-off fields greater than 3 kV/cm. Similarly, carbon-filled polymer switches having a resistivity of 1 Ω-cm have already been observed (in one-shot operation) to generate inductive fields approaching 15 kV/cm after millisecond pulse conduction times. These preliminary results and the demand for high power repetitive pulsers may spur development of materials that will be better suited for many applications, perhaps including inductive storage repetitive output pulsers.

Thorough characterization of change-of-state materials already in use for low power applications described here and those yet to be developed for appropriately high level of current handling, switching speeds and voltage hold-off will lead, it is hoped, to high energy pulsed power applications. For example, the concerted effort to develop conducting polymers to replace copper in electrical equipment and in distribution lines, in order to open a way to cheaper electrical circuits and to reduce the weight of the equipment, has already led to the development and fabrication of highly conducting polyacetylene and polypyrrole[25, 26]. These materials, in contrast to those described in this chapter, derive their low resistivity (10^{-3} Ω-cm and 10^{-2} Ω-cm respectively) from the special arrangement of double bonds within each of the polymer chains and by the addition of electron acceptors such as iodine, or donors, such as lithium. This arrangement is fragile and the material becomes unstable, losing its conductivity at temperatures near or somewhat above 100°C. Although the correlation between heating rate and decrease of conductivity has not yet been investigated, these materials are likely candidates for inexpensive throwaway high power opening switches that will operate similarly to fuses, without, however, generating strong explosive forces associated with fuse vaporization.

REFERENCES

1. R. D. Ford, I. M. Vitkovitsky, and M. Kahn, "Application of Non-Linear Resistors to Inductive Switching," special issue on electrical insulation, *Trans. Electrical Insulation* **EI-20**, 29–37 (Feb. 1985).
2. R. D. Ford and I. M. Vitkovitsky, "Nonlinear Resistive Switches," *Proceedings of the Workshop on Solid State Switches for Pulsed Power,* W. M. Portnoy and M. Kristiansen, Eds., Texas Tech University, Lubbock, TX, pp. 309–326 (1983).
3. W. C. Nunnally and R. B. Hammond, "80 MW Photoconductor Power Switch," *Appl. Phys. Lett.* **44** (1984). Useful material is also provided by W. C. Nunnally, "Photoconductive Pulse Power Switches," *Digest of Technical Papers of the Fourth IEEE Pulse Power Conference,* Albuquerque, NM (M. F. Rose and T. H. Martin, Eds.), IEEE Cat. No. 83CH1908-3, pp. 620–623 (1983).
4. R. J. Commisso, R. F. Fernsler, V. E. Scherrer, and I. M. Vitkovitsky, "High-Power Electron-Beam Controlled Switches," *Rev. Sci. Inst.* **55**, 1834–1840 (1984).
5. M. Kahn, "Effects of Heat Treatment on the PTCR Anomaly in Semiconducting Ba-TiO₃," *The American Ceramic Society* **50**, 676–680 (1971).
6. A. F. Doljak, "Polyswitch PTC Devices-A New Low Resistance Conductive Polymer-Based PTC Device for Overcurrent Protection," *IEEE Trans. on Components, Hybrids and Manufacturing Technology,* **CHMT-4**, 372 (1981).

7. V. K. Mathur, C. S. Chang, Wei-Lou Cao, M. J. Rhee, and Chi H. Lee, "Multikilovolt Picosecond Optoelectronic Switching in $CdS_{0.5}Se_{0.5}$," *IEEE Journal of Quantum Electronics,* **QE-18,** 205–209 (1982).

8. R. B. Hammond, "Extremely Fast Rise-Time Switches", op. cit. Ref. 2, pp. 222–257.

9. G. Morou, W. Knox, S. Williamson, "Picosecond High Power Switching and Applications," *Laser Focus,* 97 (April 1982).

10. RAYCHEM Corp. Bulletins, "Polyswitch Products," Series TEM-16K, TEM019K and BSR20528, Menlo Park, CA 94025 (1982).

11. Y. Shapiro, S. Foner, and T. B. Reed, "EuO. I. Resistivity and Hall Effect in Fields up to 150 kOe." *Phys. Rev. B,* **8,** 2299–2315 (1973).

12. V. A. Glukhikh, A. I. Kostenko, N. A. Monoszon, V. A. Tishchenko, and G. V. Trokhachev, "Results of Investigations of High Specific Breaking Power Superconducting Switches," *Proceedings of the Seventh Symposium on the Engineering Problems of Fusion Research,* Knoxville, TN, IEEE Pub. No. 77CH1267-4-NPS, pp. 912–915 (1977).

13. K. H. Schoenbach, M. Kristiansen, G. Schaefer, "A Review of Opening Switch Technology for Inductive Energy Storage," *Proc. of IEEE,* **72,** 1019–1040 (1984).

14. D. U. Gubser, "High Power Superconducting Switches," op. cit. Ref. 2, pp. 361–362.

15. D. A. Kleinman and A. L. Schawlow, "Corbino Disk," *J. Appl. Phys.* **31,** 2176–2187 (1960). More recent work showing opening times of 5 μs is E. K. Inall, A. E. Robson, and P. J. Turchi, "Application of the Hall Effect to the Switching of Inductive Circuits," *Rev. Sci. Inst.* **48,** p. 462–463 (1977).

16. S. R. Ovshinsky, U.S. Patent 3271584 (1966).

17. I. Ueda, M. Takiuchi, S. Ikegami, H. Sato, "Dielectric Breakdown of Polycrystalline $BaTiO_3$," *J. Phys. Soc. of Japan* **19,** 1267 (1964).

18. D. H. Auston, "Picosecond Optoelectronic Switching and Gating in Silicon," *Appl. Phys. Lett.* **26,** 101 (1975).

19. C. H. Lee, ed., *Picosecond Optoelectronic Devices,* Academic Press, New York (1984).

20. O. F. Zucker, J. R. Long, V. L. Smith, D. J. Page, and J. S. Roberts, "Nanosecond Switching of High Power Laser Activated Silicon Switches," Lawrence Livermore Laboratory Report UCRL-77449 (1975).

21. R. D. Ford and I. M. Vitkovitsky, "Inductive Storage Pulse-Train Generator," *IEEE Trans. on Electron Devices* **ED-26,** 1527–1531 (1979).

22. A. Robson, Naval Research Laboratory, Washington, DC, private communication (1983).

23. Sprague Company Engineering Bulletin, 7701, Sprague Electric Co., 149 Marshall St., North Adams, MA 01247 (1974).

24. R. D. Ford, D. J. Jenkins, W. H. Lupton, and I. M. Vitkovitsky, "Multi-Megajoule Inductive Storage for Particle Beam Production and Plasma Implosions," *Proceedings of the Fourth International Topical Conference on High-Power Electron and Ion-Beam Research and Technology,* H. J. Doucet and J. M. Buzzi, eds. Ecole Polytechnique, Palaiseau, France, pp. 743–950 (1981).

25. "Scientists Solve the Problems of Plastic Conductors," *New Scientist* **103,** 22 (30 Aug. 1984).

26. R. L. Greene and G. B. Street, "Conducting Organic Materials," *Science* **226,** 651–656 (1984).

Appendix
Shielding and Safety Procedures

BACKGROUND

Application of high power switches in laboratory and in operational pulsed power systems is associated with two problems that must be of concern to the user of such equipment. These problems are electromagnetic pulse interference and safety hazards. While the first problem can cause serious measuring errors, interfere with neighboring electronic equipment, and even lead to faulty operation of the device which uses such switches, the second problem—the safety of personnel—warrants very careful attention during the design phase, as well as during operation. These two problems are not totally unrelated. It is, for example, possible for the electromagnetic fields associated with a functioning pulse power equipment to couple sufficiently strongly to surrounding conductors to cause powerful currents that lead to dangerous effects such as sparking or high voltage shocks.

To demonstrate the potential for disaster, it is worthwhile to relate an accident in the 1970's at the Naval Research Laboratory. A large pulsed homopolar generator (shown in Fig. 2–5) was being used to test current interruption switches being developed for generating pulsed output from inductive storage at levels of more than 100 kV. Inadvertently, a steel carriage on metal wheels used in the maintenance of the generator was left nearby, perhaps one meter away from the generator, as the homopolar generator discs were spun up to the speed required by the test. As the inertial energy of the discs was converted to magnetic energy by the self-excited multiturn solenoid of the homopolar generator, the unshielded magnetic field set the carriage weighing hundreds of kilograms in motion. The metal carriage wheels, rolling over the pressurized hydraulic fluid hoses, cut the hoses, causing the high operating pressure to form a cloud of fluid mist. This failure was insufficient to terminate the test automatically. The switch functioned properly and generated high voltage output pulse that in turn led to some sparking. The spark ignited the mist producing an extremely fast fire—nearly an explosion. The operators in the adjoining room were protected by a heavy wall from being engulfed by the fire. However, the rapid rise in pressure of the overheated air blew the door open and caused, fortunately, only relatively mild flash burns on the skin of the personnel facing the door. Major investigation of this incident led to many recommendations for changes in experimental layouts and in procedures. The basis for these recommendations are the fundamentals of safety practices discussed later in this chapter. Specifically, analysis of the accident highlighted the hazards posed by flammable liquids in the high voltage environment. The general implication is that, because they are used throughout pulsed power technology, it is necessary to consider such liquids and indeed other system components as acting in synergism to properly assess the hazards associated with high voltage, and hence to foster personnel safety and prevent equipment damage.

INTERFERENCE

General Problem

In analogy to *continuous* sources of electromagnetic noise such as that from electrical power lines, automobile electrical systems, dental x-ray machines, and a host of other equipment found in everyday use, *pulsed* power systems generate electromagnetic noise in a variety of ways. The noise emanating from pulsed systems differs from that generated by continuously operating systems. It is limited in duration to the characteristic time of the pulses generated in various parts of the pulser. Very high power pulse output, in comparison with the much smaller output power of cw systems, can lead to very severe disturbances in the equipment associated with the pulser such as recording oscilloscopes, pulser control consoles with their arrays of delay generators or sequencers, trigger units, and other components. The effects can be comparable to or exceed those of natural lightning when it couples to power lines or other circuits or when its radiation causes "static." The reason that pulsed power sources intefere most disruptively with the equipment associated with these sources is that pulsed systems, especially those producing single-pulse output, produce electromagnetic (EM) noise lasting a very short span of time. Thus, the probability of the occurrence of an undesirable pulse event at an instant to interfere with other unrelated equipment is small and it is significant only in relation to the equipment associated with the pulser itself. For example, the consecutive triggering of multiple-stage opening switches may be easily subverted if the first stage switch generates sufficient noise to couple directly into the trigger unit for the second stage, bypassing the command pulse delay unit set for appropriate timing of the second stage switch and causing it to prefire. Unlike the case of natural lightning, the source of the EM noise is in close proximity to the systems vulnerable to such noise. It is often necessary to shield carefully the source of the noise as well as the susceptible equipment.

Noise Sources and Coupling

The two main sources of electromagnetic noise are (1) the current loops constituting the pulser and (2) the capacitive coupling associated with surfaces charged to high voltage relative to ground. Current loops are interspersed with switches designed to operate in a manner that shapes output pulses into desired waveforms. As a switch is activated to initiate or terminate the current in a loop, its action also causes changes in the voltage at various points in the pulser. The resulting change in the amplitude of the electric field induces waves on nearby conductors. Secondary sources of EM noise are various contact points that spark when they carry high current. Being localized, the noise frequency from such sparks usually is much higher then the fundamental frequency of the pulse being generated.

Methods for reducing electromagnetic interference, or pick-up, from high power pulsers follow the general practices developed for dealing with cw and pulsed low power systems. As J. C. Martin points out[1], however, some aspects of this problem are more accentuated since often the instrumentation uses milliamperes for monitoring the response, for example, of strain gauges, in the presence of currents from low voltage capacitor banks that exceed 10^7 A. On the other hand, methods for protection of the equipment in the vicinity of high power pulsers against stray pick-up that can cause burnout of electronic components and other damage have been worked out in conjunction with the studies of the effects of the electromagnetic pulse (EMP) generated by the explosion of a nuclear weapon. One of the more comprehensive studies that includes the discussion of protection techniques and lists firms manufacturing protection components is the Bell Laboratories Electrical Protection Department report[2], which in 1983 went into its fifth and latest printing.

Pick-up of electrogmagnetic noise occurs as a result of inductive or capacitive coupling. Although, often many interacting loops, formed by various connections (including capacitive coupling or displacement currents) between the pulser, the external ac network, and the signal cables are too difficult to analyze completely, simplification of such loops, shielding techniques, and decoupling using recently developed optical signal lines allow most of the interference problems to be adequately managed.

Of the two types of coupling, inductive coupling is the more difficult to deal with because of the large magnetic fields associated with the high currents found in high power pulsers. The magnetic field, B, generated around a conductor carrying current I is, in mks units:

$$B = \frac{\mu_0 I}{2\pi R} \quad \text{(W/m}^2\text{)} \tag{A-1}$$

where $\mu_0 = 4\pi \times 10^{-7}$ H/m for free space and R is the distance from the center of the conductor. (Eq. A-1 can be written in units useful for laboratory practice as $B = I/5R$ where B is in gauss, I is in amperes and R is in centimeters). The expression for B can then be used with Faraday's law to obtain the induced voltage V, in a loop of area A normal to the magnetic flux $\phi = \int BdA$:

$$V = \frac{d\phi}{dt} = \frac{d}{dt} \int BdA \quad \text{(V)} \tag{A-2}$$

If the Area A is given by πr^2, then assuming that B is constant over A, equations (A-1) and (A-2) give

$$V = \frac{d}{dt} \left(\frac{\mu_0 I}{2\pi R} \right) (\pi r^2) = \mu_0 \frac{r^2}{2R} \frac{dI}{dt} \quad \text{(V)} \tag{A-3}$$

Equation A-3 suggests that the distance, R, from the current lead to conductors subjected to induced pick-up noise should be large if the noise amplitude is to be reduced. Because this is not possible in many cases, as, for example, when voltmeter probe leads are attached across the load, the pick-up area should be minimized (i.e., the probe leads should run together).

A numerical example illustrates the magnitude of the pick-up problem. An 1 MA current rising in 1 μs produces $dI/dt = 10^{12}$ A/s. The changing magnetic field intercepting a loop of 1 cm diameter at a distance of 10 cm from the current conductor will, thus, induce about a 300 V signal in the loop, according to EQ. A-3. The degree of interference caused by such a signal depends, of course, on the path connecting such a loop to the recording equipment.

Electrostatic coupling occurs when signal leads, perhaps the same voltmeter probe leads mentioned earlier, are in the vicinity of an electrode with potential V_e with respect to ground, which increases (or decreases) in time t. If the lead is isolated from the ground, there will be induced a charge q due to the stray capacity, C_s, between the electrode and the isolated lead.

Since $q = C_s V_s$, the rate of rise of the charge (i.e., the displacement current, i, required to charge the stray capacity) is given by

$$\frac{dq}{dt} = C_s \frac{dV_s}{dt} \quad \text{(A)} \qquad\qquad \text{(A-4)}$$

If the lead is not totally isolated but connected to ground via a resistance, R_0, such as oscilloscope input impedance, the induced voltage signal across R_0 could be as high as

$$R_0 \frac{dq}{dt} = R_0 C_s \frac{dV_s}{dt} \quad \text{(V)} \qquad\qquad \text{(A-5)}$$

if R_0 is such that the characteristic time, $R_0 C_s$, is large compared to t.

To get an estimate of the value of C_s, the geometry associated with the high voltage and the diagnostic or other representative lead surfaces can be approximated by considering the spacing, s, between them and making some assumption about the effective areas contributing to the stray capacity. An order of magnitude estimate results when the area, A, is taken to be that given by the product of the lead diameter and the length of the lead being equal to the separation, s, i.e., in mks units:

$$C_s \simeq 4\pi k_0 \, \epsilon \, A/s \quad \text{(F)} \qquad\qquad \text{(A-6)}$$

where $k_0 = \frac{1}{36}\pi \times 10^9$ F/m is the permittivity of free space and ϵ is the dielectric constant of the medium associated with the stray capacity, C_s.

The magnitude of the electrostatic pick-up in pulse power equipment can be evaluated, for example, for a probe lead 10 cm away from the switch electrode. If the lead has 1 cm diameter and runs parallel to the electrode surface, the stray capacity is of the order of 10^{-12} F (1 pF). If the resistance of the lead to ground, R_0, is such that $R_0 C_s \gg t$, then the noise voltage, V_s will rise to full value V_e. Normally, diagnostic leads are connected to ground via 50 Ω terminations at the oscilloscope (i.e., $R_0 = 50 \, \Omega$). Thus, $R_0 C_s$ time is short ($\sim 5 \times 10^{-11}$ s). This will reduce the noise amplitude, V_s, to a level given by the ratio $(R_0 C_s)/t$, i.e., $V_s = V_e R_0 C_s/t$. For $V_e = 1$ MV signal on the electrode of a pulser rising in 10^{-7} s, the induced noise across the scope will be 500 V. Another reduction of the noise level will occur from any capacitive coupling of the lead to ground. An order of magnitude estimate of the induced noise signal reduction can be obtained by calculating the capacitive division of the signal voltage by the stray capacities to the high voltage point and to the ground plane. Eq. A-6 should be used for calculating the stray capacity of the lead to ground. Since often the ground plane is large (e.g., forming an enclosure for the pulse power source), this re-duction could be quite substantial.

The frequency of pick-up noise discussed so far is related to the natural frequencies of the pulsed power system and to the existence of various stray capacities. A source of noise with frequencies much higher than the natural frequencies also is generated in poor metal-to-metal contacts that spark when high current passes through them. The high frequencies are associated with the capacities formed by the gaps between contact surfaces. Well de-signed switches have metal-to-metal contacts under sufficiently high pres-sure to avoid sparking. Nevertheless, it can be expected that as a switch continues in operation, its electrodes and connections to high current leads will deteriorate, eventually leading to noise.

There are other sources of noise that are not directly related to switch or pulser operation. These are related to poor signal cables or to bad joints in the construction of shielding, allowing pulser-induced ground currents to penetrate into shielded enclosures. These can be recognized by the presence of ample high frequency components superimposed on the measurement waveforms.

Finally, because high power switching technology frequently deals with short pulses, steep wavefronts and large signals, it is necessary to remember that all signal *reflections* should be eliminated or controlled in such a way that they do not interfere with diagnostic measurements.

Methods of Shielding

To prevent stray electric and magnetic fields from interfering with the op-eration of pulsers and of diagnostic measurements, measuring and control equipment is enclosed in metal boxes called *Faraday cages.* Circuits com-pletely within perfectly conducting boxes cannot be affected by the steady or changing electromagnetic fields outside, and electromagnetic fields gen-erated inside such enclosures cannot penetrate to the outside.

Because the shielding enclosures are not ideal and contain openings for ventilation and for power cables, and have leaking discontinuities produced in their fabrication, it is a good practice to enclose as completely as possible the pulser elements, including switches, to minimize the amplitude of ground currents and of the radiating noise.

In the cases in which both the pulser and the measuring equipment are shielded by separate enclosures, the control and diagnostic signals fre-quently are also connected by cables with their outer skin formed by solid metal tubes. An idealized arrangement is shown in Fig. A–1. There are a host of practical problems which arise from imperfect shielding. First is the finite conductivity of the enclosure metal. This means that for frequencies

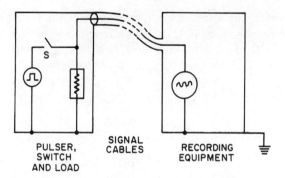

Fig. A-1. Practical shielding approach for high noise-environment, using two separate metal shield rooms (or boxes). The first enclosure prevents the emanation of noise from the pulsed power installation, while the other attenuates any leakage before it reaches the recording apparatus. The control and diagnostic signal lines are also completely enclosed to avoid breaks in the shielding.

below those determined by the skin thickness associated with the metal's conductivity, the fields do penetrate into the enclosure. Thus such measuring devices as photomultipliers, which are very sensitive to magnetic fields, cannot be operated near the low frequency magnetic fields of large solenoids unless very thick shielding is provided. Second, various penetrations allow currents induced on the outside of shields to diffuse inside and couple to sensitive equipment either through grounding connections or through capacitive coupling between the equipment components and the inside walls of the shield.

Solid coaxial cables that serve to connect shielded enclosures shown in Fig. A-1, as well as to extend the shielded enclosure to various transducers, are available with braided and solid outer jackets. The solid jacket cables provide much better protection against pick-up. Coaxial cables with approprite connectors are produced by several manufacturers in the United States and in other countries. Most common is the 50 Ω cable. Cables produced, for example, by Cablewave Systems range from 1.25 cm in diameter (designated HCC12-50J) to 8.87 cm in diameter (designated HCC312-50J)[3].

To minimize the secondary coupling induced by the presence of shield penetrations, it is necessary to analyze the specific configuration of the pulser, diagnostics equipment, and the various connecting cables. In the analysis, the resistance and inductance of long cables and grounding leads should be accounted for. Some limitations on cable length are also imposed by the cables' frequency response. (As the signal frequency rises from 1 MHz to 1 GHz, the attenuation per 100 m of cable increases from 0.32 db for Cablewave Systems' HCC12-50J. The larger cables have less severe attenuation. For example, the attenuation in db/100 m of the HCC312-50J cable is ten times lower than the HC12-50J cable over the 1 MHz to 1 GHz range of frequencies.) Often unsuspected capacitances also permit reactive currents to flow in various lines. Judicious use of filters, isolating transformers, and grounding resistors can minimize these problems. Such techniques are discussed in many sources, including Ref. 2, 4, and 5.

Other Methods of Reducing Interference

Less direct methods, just mentioned, for dealing with electromagnetic inteference recognize that some noise is unavoidable and manipulate its relationship to the diagnostic or control signal. One approach uses the signal travel time in a cable connecting the transducer and a recording device (e.g., an oscilloscope) to delay the diagnostic signal relative to the arrival time of the noise signal to the recording device via other paths such as a ground loop or direct radiation. Sufficient length of signal cables, such as 50 Ω cables, can be used to produce delays of 1 to 2 μs. For longer delays, other delay devices can be used. Another approach uses balanced input into a differential amplifier with common mode rejection.

With the development of optical fibers for wide band signal transmission, it is possible to break completely any ground current loops as well as to remove signal lines from regions of strong electric and magnetic fields. This approach eliminates not only many of the interference problems but also allows the placement of probes into regions of the pulser that are inaccessible to metallic conductors because of the effects of the latter on pulser performance, as would be the case, for example, if near-breakdown electric fields were enhanced by the presence of grounded conductors. Optoelectronic devices and optical fibers have been used in conjunction with triggering of the class of explosively driven switches discussed in Chapter 8. These devices (FS40 control and firing modules and Series 188 optical isolators) have been made available by Reynolds Industries Systems[6]. Similar less rugged devices are becoming available to replace traditional signal generators and cables[7].

SAFETY

Along with the broadening of electrical power applications in modern society, very high power equipment has proliferated in industry and in laboratories in the last several decades. The safety standards associated with the use of electricity from an earlier period has required significant extension to account for changes in equipment operation, appearance of new materials, and the significant extension of the current, voltage, and pulse parameters required in research and advanced technologies. Thus, various new studies and guides have been published to provide better protection for the personnel working with or affected by high power equipment[8, 9, 10, 11].

Hazards

There are several types of hazards associated with the operation of high power switches that derive not only from the presence of high voltages but also high energy density. In addition to the usual danger of electric shock, the presence of strong electric fields, x-ray radiation[12], chemical hazards[12, 13] and fire contribute their own perils. Further, because of the large energy and energy density handled by high power switches, mechanical forces and the potential for explosive-like failures constitute major concerns in safely operating such switches.

Electric Shock.   Each year over 1000 people are electrocuted in the United States and many more suffer from injuries caused by the electric shock resulting from contact with energized parts of electrical equipment or wires[9]. Most of these fatalities and injuries occur in exposures to the so-called "low" voltages, especially those of the 110–120 V systems. Pulsed systems differ from these hazards in two significant ways. Often the voltages are very high and pulse durations are short. Very high voltages, i.e., above tens of kilovolts, lead to breakdown discharges, such as surface flashover, which go beyond the common experience that people have regarding electrical insulation. The special physiological effects of short duration pulses are discussed, in References 2, 8, and 14.

Generally shock hazards are classified according to current through the human body[14] or the amount of energy deposited in the body[2]. Voltage is important only in that it enables the current to pass through the body, as happens when breakdown through dry skin occurs, and deposit the incapaciting energy. (The internal resistance of the human body is about 100 Ω between ears and 500 Ω from a hand to a foot. The skin is important in increasing the resistance to current when dry, but is much less significant when wet. The ratio of dry-to-wet skin resistance is about 1000).

Appendix C of Ref. 7 emphasizes that the ability of human body to withstand electric energy is so miniscule that most electrical systems are hazardous to health and life. It continues with a list of the effects on the body resulting from decreasing shock strength. Thus, 100 mA ac, 500 mA dc, or 50 J impulse discharges through the body are considered lethal. Shock is relatively severe at 20 mA, when breathing becomes labored; breathing ceases completely at less then 75 mA. At currents higher then 200 mA, muscular contractions are so severe that the heart is forcibly clamped, which actually protects it from going into venticular fibrillation so that the chances of survival are still good. Equipment capable of driving 10 mA ac or 60 mA dc through a 500 Ω load or having 0.25 J of stored energy presents a serious shock hazard. Minor shock can also cause serious injury due to reflex body action. Electrodes delivering energy at 0.8 J/cm² produce minor burns on the skin. Ref. 2 quotes similar conclusions with a warning that the biological effects from microsecond pulses have not been extensively studied, and the limited studies have been confined to single (not repetitive) pulses.

Intense Electromagnetic Fields. Exposure to magnetic and electric fields associated with high power switch operation does not appear to cause problems. (The effects of single pulse short duration fields on the human body are considered in Ref. 2). However, as such fields are applied repetitively to the human body, it is possible that harmful effects can be induced[15]. Radiation at wavelengths that allow these fields to penetrate deep into the body can lead to harmful effects even when the full-body temperature rise is relatively small. Reference 2 suggests that a 10°C transient rise in body temperature can be dangerous and a sustained rise of 1°C can be fatal. The possibility of the latter hazard arises from the poor thermal dissipation of heat from the internal parts of the body which may be overheated[15]. The authors[2, 15] point out, however, that, based on RF safety standards of the United States of America Standards Institute, it would be difficult to get sufficient whole-body exposure from single short pulses to effect serious injury.

X-Rays. In switches operating at voltages exceeding tens of kilovolts, especially those used for megavolt operation, there exists a danger of x-ray formation. This can happen especially in low pressure or vacuum switches due to acceleration of electrons by applied electric fields toward positive electrodes. (This is a similar hazard to that encountered in working with high voltage power supply tubes, i.e., operating at voltages higher then 15 kV[12]). If the mean free path for electrons is greater than that of the electrode spacing, the electrons (at least during some phase of switching) could gain energy from the applied field while traversing the spacing. As the electrons are intercepted by the anode, part of their energy will be converted to x-rays. The amount converted increases linearly with the nuclear charge, Z, of the target material and with the electron energy, U_e. One MeV electrons stopped in intermediate Z materials such as copper convert to x-rays with about 0.5% efficiency. This efficiency changes approximately in proportion to U_e^2. The radiation consists mostly of bremsstrahlung and K and L lines. For thin targets, i.e., targets transparent to most of the radiation, the peak of the bremsstrahlung spectum is at an energy of about one-third of the maximum photon energy (which is equal to the incident electron energy). In the worst case of high switch currents (in the megampere range) and high voltages, thin unshielded target electrodes can produce a radiation dose in the range of several rads at a distance of a few meters, much more than the safe allowable dose.

Chemical Toxicity.   Chemical reactions of switch dielectric materials occur during discharges or as a result of the corona surrounding high voltage electrodes. These reactions may occur in the switches themselves, where discharges, electron beams, or radiation can change the composition of switch materials or that of ancillary equipment. This possibility has been pointed out in conjunction with laser operations and laser irradiation of materials[13], which resemble conditions encountered in high power switches. Decomposition products from these reactions can form toxic substances in post-discharge recombinations or in reactions with chemicals in the environment. Ozone may be a product of such reactions when simple air gap switches are used (and entrapped ozone can explode when shocked[13]). (Brand[16] has attempted to relate the toxic properties of insulating gases to other characteristics of these gases to provide some guides regarding potential biological hazards.) Those higher performance switches that use SF_6 as the dielectric produce a large variety of reactants, some of which can be toxic.[17] Simple ventilation techniques usually are sufficient for dealing with chemical hazards.

Mechanical Failures.   The mechanical hazards associated with high power switches are related to many different causes. Some switches, such as those described in Chapter 7, rely on a substantial amount of explosives to perform their function. Other switches are pressurized and must withstand a transient pressure increase induced by the absorption of electrical energy during switching. Still other switches handle strong magnetic pressures. Material failure or improper asembly can easily cause what literally is an explosion of the switch housing, electrodes, and connecting leads. The energy for generating the explosion can come from external chemical sources, such as the explosive cords, or from the potential energy associated with the large gas pressures used in other types of switches, as well as from the input of electrical energy, especially if it is misdirected from other switches that normally share the switching burden in a pulser.

Guidelines

There are well-established precautionary practices that have been adapted[11] for dealing safely with pulsed power equipment used in laser research and its application. These practices should be extended to all high power switching applications and supplemented by a clear understanding of the other hazards described above.

Reference 11 provides a set of guidelines for safe handling of high voltage equipment, calling for the removal of all personal metallic objects (such as rings, metallic watch bands and necklaces), avoiding the wetting of skin and clothing, and avoiding the use of both hands when manipulating potentially high voltage components. With high voltages, all floors should be regarded as good ground and as conducting, unless they are covered with dry rubber mats suitable for electrical work. Pulsed power equipment should be provided with emergency shut-off switches. Grounded metal covers with interlocks should enclose high power equipment to prevent accidental contact with current-carrying conductors. When personnel approach such equipment, automatic and manual crowbar (dump) resistors should be engaged and the action confirmed visibly before any access parts are opened.

Finally, no one should work alone with equipment that poses electrical shock hazard (and those who rely on the pacemaker should avoid areas of high pulsed fields[2], especially when the equipment is operated in repetitive mode).

When, in addition to electric hazards, there also exists a potential for dangerous mechanical failures, blast shields, concrete walls, or other suitable protection (including large separation) should be imposed between the operator and the equipment during its operation. The extent of damage caused by mechanical failure can be appreciated via a comparison of suddenly released electrical energy, resulting from such a failure with the chemical energy of explosives: the energy content of 1 kilogram of TNT is equivalent to 4 MJ of energy. There are many pulsed power sources storing energy at a megajoule level. A switch failure can release this energy explosively and cause substantial damage. It should be remembered that hand grenades, which can be devastating, contain only a fraction of a kilogram of an explosive.

Respect of the hazards and thoughtfulness in dealing with them can save lives.

REFERENCES

1. J. C. Martin, "Nanosecond Pulse Techniques," *Pulsed Electrical Power Circuit and Electromagnetic System Design Notes,* Air Force Weapons Laboratory Report AFWL-TR-73-166, Sect. IV (1973).
2. R. Sherman, R. A. De Moss, W. C. Freeman, G. J. Greco, D. G. Larson, L. Levey, and D. S. Wilson, "EMP Engineering and Design Principles," Technical Publication Department, Bell Laboratories, Whippany, NJ (1983—fifth printing). An extensive bibliography dealing with the electrical hazards is provided here.

3. Cablewave Systems, Catalog 500A, pp. 68–115, Cableway Systems, New Haven, Connecticut 06473 (1979). For additional information on cables and connectors, the following catalogues, Andrew Corp., Orlando, Illinois, "Antenna Systems Catalog 30," (1978); and Alpha Wire Corp., Elizabeth, NJ 07207, Catalog W-M-8 (1973), can be consulted.

4. P. R. Barnes, E. F. Vance, and H. W. Askins, "Nuclear Electromagnetic Pulse (EMP) and Electric Power Systems," Oak Ridge National Laboratory Report ORNL-6033, pp. 40–43, (1984).

5. R. Morrison, "Basic Considerations - Shielding and Grounding for Instrumentation Systems," Dynamics Instrumentation Co. Report TP/755-1 (undated). Comprehensive discussion of shielding methods is given in the following: J. Wiesinger, "Basic Principles of Grounding and Shielding with Respect to Equivalent Circuits," *Fast Electrical and Optical Measurements,* Vol. 1, J. E. Thompson and L. H. Luessen, eds., Martinus Nijhoff Publishers, Boston MA, (1986).

6. Reynolds Industries Systems, Inc., Catalogue, p. 5, RISI, Los Angeles, CA 90066.

7. Mazwell Laboratories, Inc. Data Sheet for Fiber Optic Trigger Link, Model 40290, undated; MLI, 8888 Balboa Ave., San Diego, CA 92123. This device operates on 450 V input and provide 50 V output into a 50 Ω load. Fiber optic cables are available in lengths up to 500 meters.

8. U.S. Atomic Energy Commission Safety and Fire Protection Technical Bulletin No. 13, "Electrical Safety and Guides for Research," U.S. AEC., Division of Operational Safety, Washington, DC (1967). Specifically, Chapter II of this report is very useful in dealing with safety issues associated with pulsed power equipment: capacitors, equipment enclosures, inductors and coils, resistors, switches, and instrumentation and control.

9. "NRL Safety and Occupational Health Manual," Naval Research Laboratory Report NRLINST 5100.13A, Ch. 11 (1980).

10. F. E. McElroy, ed., "Accident Prevention Manual for Industrial Operations, Engineering and Technology," 8th Ed., published by National Safety Council, Ch. 15, (1980).

11. O. Sliney and M. Wolbarsht, *Safety with Lasers and Other Optical Sources,* Plenum, New York, Ch. 28 (1980).

12. "Guide for Laser Installations" *National Safety News,* 85–87 (Dec. 1969).

13. D. MacKeen, S. Fine, and E. Klein, "Safety Note: Toxic and Explosive Hazards Associated with Lasers," *Laser Focus,* 47–49 (Oct. 1968).

14. C. F. Dalziel, "Dangerous Electric Currents," *Trans. AIEE* **65,** 579–585 (1946).

15. W. W. Mumford, "Some Technical Aspects of Microwave Hazards," Proceedings of IRE **49,** 427–447 (1961).

16. K. P. Brand, "Dielectric Strength, Boiling Point and Toxicity of Gases—Different Aspects of the Same Basic Molecular Properties," *IEEE Trans. on Electrical Insulation* **EI-17,** 451–455 (1982).

17. F. J. J. G. Jansen, "Decomposition of SF_6 by Arc Discharges and the Determination of the Reaction Products," *Program and Abstracts of the Fifth International Symposium on Gaseous Dielectrics,* Oak Ridge National Laboratory, Oak Ridge, TN, p. 9 (May 1987).

Index